SpringerBriefs in Stem Cells

For further volumes:
http://www.springer.com/series/10206

Sibel Yildirim

Induced Pluripotent Stem Cells

Sibel Yildirim PhD
Department of Pediatric Dentistry
Faculty of Dentistry
Selcuk University
Konya
Turkey
e-mail: ysibel@gmail.com

ISSN 2192-8118
e-ISSN 2192-8126
ISBN 978-1-4614-2205-1
e-ISBN 978-1-4614-2206-8
DOI 10.1007/978-1-4614-2206-8
Springer New York Dordrecht Heidelberg London

Library of Congress Control Number: 2011942210

Printed on acid-free paper

Springer is part of Springer Science+Business Media (www.springer.com)

Preface

"The world is changing" is a classical concern for people over forties. However, the era we live is a little bit different than the previous generations. We are more likely in between classical and contemporary or between fictions and facts. Change is unbearably fast and the story of biology is still far from to be completed. It is only a few weeks ago (September 22, 2011) particle physicists on the particle detect or named "Oscillation Project with Emulsion-tRacking Apparatus (OPERA)" experiment detected neutrinos traveling faster than light. Although it is too early to declare Einstein's theory of special relativity is wrong, results would be said being so revolutionary at least. The same has happened in cell biology in 2006. Shinya Yamanaka and his team discovered that the completely differentiated somatic cell could exert its embryonic stem cell state potential with the available conditions. The method for turning a somatic cell to a pluripotent one was relatively easy, at least easier than one could imagine till Yamanaka's paper. They named those cells as induced pluripotent stem cells (iPSCs).

Science is used to have such sudden pulses. However, there is always a resistance to unexpected changes. It generally takes longer time than necessary to interpret the discoveries having results that are applicable to many situations. Some scientists want to directly apply those results to a daily life as quickly as possible, while the others, whom being fascinated by reaching one more level of endless mystery, display a tendency to beware there is actually more. However, some with a special consciousness embark through explorations of unknown realities.

The discovery of iPSCs brought about all. Now many researchers are trying to refine the technique to serve those cells to restore human health. Some of them are trying to reach the furthest point in the dark corridors of cell biology, while they are aware that the battery of their torch is pushing its limits. However, very few of them opened a completely renewed era in biology by mathematical biology.

This manuscript is trying to explain the fundamentals behind the iPSCs and its applications. Most importantly, it attempts to show why we have to use mathematics to go further with iPSCs or another yet undiscovered cells. The theories of Stuart Kauffman and Sui Huang pointed out the ways to solve many problems in cell biology and are being pored over by many people to quench their intellectual thirst. Dr. Huang is showing bravely how impossible to fathom by common sense of all data flooding from 'omics' works of biology. Fortunately, he is using general concepts or principles of physics and mathematics to establish a firm theoretical foundation.

Researchers from my team started from tooth regeneration and inescapably ended up with stem cell biology. I individually had begun to be involved in microbial aspects of dental diseases. Now I am being captivated by complexity and system biology, because it is hard not being exposed to the emergent patterns of every system that has a common connection: mathematics. Obviously iPSCs are providing great tools to study every aspects of biology, fundamentals through applications. In conclusion, iPSCs opened minds of scientist by showing that we should compel our imagination limits to see more.

I am grateful to Sui Huang for his generous and humble guidance. Thanks to Kursat Turksen to find me sufficient to write this manuscript. Thanks to Springer's publishing team, especially to Renata Hutter and Aleta Kalkstein for their kind assistance. Thanks also to Kamil Can Akçalı who encouraged me to follow my instincts. Last but not least thanks to Muammer Saglam for his unconditional love that led me to feel the light.

Contents

1 **Introduction** . . . 1
References . . . 2

2 **Pluripotent Cells** . . . 5
2.1 Different Pluripotent Cells . . . 5
2.2 Transcriptional Networks and Signaling Pathways of Pluripotency . . . 6
2.2.1 Transcriptional Network of Pluripotency . . . 8
References . . . 9

3 **Induced Pluripotent Stem Cells (iPSCs)** . . . 11
3.1 Generation the First iPSCs . . . 11
3.2 Reprogramming. . . . 12
3.2.1 Factor Delivery into Target Cells . . . 12
References . . . 17

4 **Molecular Mechanisms of Pluripotency**. . . . 21
4.1 Steps in Reprogramming . . . 21
4.1.1 Increase in the Cell Cycle Rate . . . 21
4.1.2 Morphological Changes . . . 22
4.1.3 Late Events Toward Pluripotency . . . 22
4.2 Mechanisms in Reprogramming . . . 23
4.2.1 Genetic Factors. . . . 23
4.2.2 Signaling Pathways . . . 24
4.3 Dynamics of Direct Reprogramming . . . 25
4.4 Epigenetic Modifications . . . 26
4.5 Similarities and Differences Between iPSCs and ESCs . . . 27
References . . . 29

5 Modeling Disease in a Dish 33
5.1 Disease-Specific iPSCs 34
5.2 Choosing Cell Sources 34
5.3 Identification of iPSC Colonies 40
5.4 Characterization of Genetic Mutation 42
5.5 iPSCs Differentiation into Desired Cell Types 42
References 44

6 Challenges to Therapeutic Potential of hiPSCs 51
6.1 Is Reprogramming Necessary for Regenerative Therapies? 53
References 55

7 New Approach to Understand the Biology of Stem Cells 57
7.1 Health Versus Disease 59
7.2 Changed Paradigm: Reprogramming as Rare But Robust Process 61
7.3 From Reductionism to Wholeness 62
7.4 More Future Considerations 66
References 68

8 Conclusion 69

About the Author 71

Index 73

Chapter 1
Introduction

Although cell fates during development are neither restrictive nor irreversible, interestingly enough the deeply rooted attitude in cell biology has been that the terminally differentiated cells have lost the potential of producing other cell types (Huang 2009). The first study showed that the nuclei of frog blastula cells could produce complete embryos when transplanted into enclosed embryos (Briggs and King 1952). Only a few years later Gurdon et al. (1958) reprogrammed fully differentiated intestinal cells from *Xenopus* by frog oocytes. It took 20 years of incubation for Evans and Kaufman to publish the successful isolation of embryonic stem cells (ESCs) in 1981 (Evans and Kaufman 1981). Although changing a somatic cell fate was achieved using frog oocytes, somatic cell nuclear transfer (NT) could not be succeeded in other species until late 1990s. Then Wilmut et al. (1997) cloned sheep Dolly. The first successful derivation of human ESCs was reported a year after (Thomson et al. 1998). However, ethical complications and paucity of human egg cells for research purposes, extremely low efficiency, high technical difficulty levels and aberrant ploidy caused researches on ESCs much more controversial than expected (Walia et al. 2011).

Earlier studies showed that the 'terminally differentiated' state of human cells was not fixed, but could be altered, and the expression of previously silent genes typical of other differentiated states could be induced (Blau and Baltimore 1991; Bhutani et al. 2010; Yamanaka and Blau 2010). Accordingly, pluripotent stem cell lines can be generated directly from completely differentiated adult somatic cells using alternative approaches, such as nuclear transfer, cell fusion and direct reprogramming. When a nucleus from a differentiated somatic cell is transplanted into an enucleated oocyte, nuclear reprogramming is initiated, leading to cloning of the original somatic cell. These experiments clearly showed that cell specialization needs only change in gene expression, not in gene content and this process of differentiation can be fully reversed (Yamanaka and Blau 2010). Heterokaryons constitute another complementary approach to nuclear reprogramming and it involves cell fusion, in which two or more cell types fuse to form single entity. It has been shown that reprogramming in heterokaryons was influenced by DNA

DOI: 10.1007/978-1-4614-2206-8_1,

methylation status, tissue of origin and the relative ratio of nuclei that dictates the balance of regulators (Blau and Baltimore 1991; Zhang et al. 2007). By using interspecies heterokaryons (mouse ESCs, human fibroblasts) Bhutani et al. (2010) have showed that mammalian AID is required for active DNA demethylation and initiation of nuclear reprogramming toward pluripotency in human somatic cells (Bhutani et al. 2010).

Intense effort to identify the master transcription factor of cell phenotypes have been exerted to support the idea that a small number of master transcription factors can control cell state in various cell types including reprogramming and trans-differentiation (Young 2011). Only in 2006, attempts to identify the main regulators of the ESC state was reached to a plateau with the study of Shinya Yamanaka and colleagues who showed that a combination of only four transcription factors could generate ESC-like pluripotent cells from mouse fibroblasts (Takahashi and Yamanaka 2006). These generated cells are called induced pluripotent stem cells (iPSCs). The discovery of factor-directed reprogramming had a seismic effect on stem cell biology and its potential application (Wilmut et al. 2011). Today many fundamental systems in biology are changing to accept that mature body cells could be reverted to an embryonic state without the help of eggs or embryos.

Thence, reprogramming not only sidesteps the necessity of using embryos to collect and culture ESCs, but also comes with the additional expected advantage of circumventing the immunological problems associated with engraftment, which includes transplant rejection and graft versus host disease. On the other hand the efficacy of the technique is just a wide place in the road. There are still many technical roadblocks in the process. This review will focus on the story of iPSCs that opened a new era in cell biology only in the very beginning of 2000.

References

Bhutani N et al (2010) Reprogramming towards pluripotency requires AID-dependent DNA demethylation. Nature 463(7284):1042–1047

Blau HM, Baltimore D (1991) Differentiation requires continuous regulation. J Cell Biol 112(5):781–783

Briggs R, King TJ (1952) Transplantation of Living Nuclei From Blastula Cells into Enucleated Frogs' Eggs. Proc Natl Acad Sci USA 38(5):455–463

Evans MJ, Kaufman MH (1981) Establishment in culture of pluripotential cells from mouse embryos. Nature 292(5819):154–156

Gurdon JB et al (1958) Sexually mature individuals of *Xenopus laevis* from the transplantation of single somatic nuclei. Nature 182(4627):64–65

Huang S (2009) Reprogramming cell fates: reconciling rarity with robustness. Bioessays 31(5):546–560

Takahashi K, Yamanaka S (2006) Induction of pluripotent stem cells from mouse embryonic and adult fibroblast cultures by defined factors. Cell 126(4):663–676

Thomson JA et al (1998) Embryonic stem cell lines derived from human blastocysts. Science 282(5391):1145–1147

Walia B et al (2011) Induced pluripotent stem cells: fundamentals and applications of the reprogramming process and its ramifications on regenerative medicine. Stem Cell Rev Jun 14. [Epub ahead of print]

Wilmut I et al (1997) Viable offspring derived from fetal and adult mammalian cells. Nature 385(6619):810–813

Wilmut I et al (2011) The evolving biology of cell reprogramming. Philos Trans R Soc Lond B Biol Sci 366(1575):2183–2197

Yamanaka S, Blau HM (2010) Nuclear reprogramming to a pluripotent state by three approaches. Nature 465(7299):704–712

Young RA (2011) Control of the embryonic stem cell state. Cell 144(6):940–954

Zhang F et al (2007) Active tissue-specific DNA demethylation conferred by somatic cell nuclei in stable heterokaryons. Proc Natl Acad Sci USA 104(11):4395–4400

Chapter 2
Pluripotent Cells

2.1 Different Pluripotent Cells

The defining properties of ESCs are the ability to proliferate indefinitely without commitment to any cell lineages (self-renewal) and the capacity to differentiate into cell lineages from three germ layers (pluripotency) (Evans and Kaufman 1981; Thomson et al. 1998). ESCs were the first pluripotent cells isolated from normal embryos derived from the inner cell mass (ICM) of preimplantation embryos (Evans and Kaufman 1981). Mouse ESCs (mESCs) contribute cells to the three germ layers and to the germline of chimeric animals when injected into mouse blastocysts. However, there are distinguishing molecular and biological characteristics between ESCs and their in vivo counterparts of the ICM. Cells of the ICM do not self-renew, and they have globally hypomethylated genome (Santos et al. 2002), whereas, ESCs have unlimited proliferation potential and they have characteristically highly methylated genome (Meissner et al. 2008).

Although the first mESC lines were derived 25 years ago using feeder-layer-based blastocyst cultures, subsequent efforts to extend the approach to other mammals have been relatively unsuccessful. Human ESCs (hESCs) could only be isolated in 1998 (Thomson et al. 1998). Mouse and human embryonic stem cells share similar features such as their ICM origin and pluripotency. On the other hand, they do have differences. The differences are related to their morphology, marker expression, transcription factor binding activities and growth factor requirements in culture conditions. mESCs depend on leukemia inhibitory factor (LIF) and bone morphogenetic protein (BMP), whereas hESCs rely on activin and fibroblast growth factor (FGF) (Thomson et al. 1998).

In 2007, two independent groups showed that pluripotent cells could also be derived from the epiblast of the implanted embryo (Brons et al. 2007; Tesar et al. 2007). Mouse epiblast stem cells (EpiSCs) are derived from the post-implantation epiblast of day 5.5 embryos in the presence of bFGF and activin (Tesar et al. 2007). Although EpiSCs are able to differentiate in vitro into the teratomas including tissues belonging to the three embryonic germ layers, they do not contribute to chimeras.

S. Yildirim, *Induced Pluripotent Stem Cells*, SpringerBriefs in Stem Cells

DOI: 10.1007/978-1-4614-2206-8_2,

Table 1 Different types of pluripotent cells and antagonistic actions of same signaling pathways on different states of pluripotency

	hESC	EpiSC	mESC
Pluripotency state	ICM-like	Post-implantation epiblast-like	ICM-like
Maturity state	Primed	Primed	Naïve
Morphology	Flattened monolayer	Flattened monolayer	Dome-shaped
Culture conditions	Activin A/bFGF	Activin A/bFGF	LIF/BMP4 2i
Pluripotency confirmation status	Teratoma formation	Teratoma formation	Tetraploid complementation Germline contribution
X chromosome inactivation	XaXi	XaXi	XaXa
BMP signaling	Induces differentiation	Induces differentiation	(+LIF) Stabilizes
TGF-β & FGF2	Support renewal	Support renewal	Induces differentiation
ERK1/2 pathway signaling	Requires	Requires	Self-renewal is enhanced by inhibition

Interestingly, hESCs share defining features with EpiSCs, yet are derived from preimplantation human embryos (Nichols and Smith 2009) (Table 1).

Pluripotent cell lines have also been derived from other embryonic and adult tissues. Embryonic germ cells (EGCs) have been derived from the primordial germ cells (PGCs) of the midgestation embryo (Matsui et al. 1992) and adult germline stem cells (male germ cells and spermatogonial stem cells) (Surani 1999) have been generated from explanted neonatal (Kanatsu-Shinohara et al. 2004) and adult (Guan et al. 2006) mouse testicular cells.

Moreover, in 2008 Chou et al. reported distinct pluripotent cells derived from blastocyst, which were defined by FGF2, activin and BIO. The authors called those cells FAB-SCs and showed that they share EpiSCs markers. However, FAB-SCs were unable to differentiate unless exposed to LIF/BMP4 (Chou et al. 2008).

With the discovery of EpiSCs, there is an emerging concept that different pluripotent states could exist, and knowledge of both transcriptional networks and signaling pathways has been vital for the precise description and dissection of the pluripotent state.

2.2 Transcriptional Networks and Signaling Pathways of Pluripotency

Different pluripotent cell types can be characterized and classified by their different growth requirements, developmental properties and pluripotency states (Table 1):

1. ICM-like pluripotent state: ESCs derived from ICM, embryonic germ cells and male germ cells or spermatogonial stem cells.
2. The post-implantation epiblast-like state: EpiSCs

These two states depend on signaling pathways that often antagonize each other (Hanna et al. 2010b). On the other hand, Nichols and Smith designated pluripotent cells as na and primed according to their maturity state of pluripotency (Nichols and Smith 2009). Isolated from the ICM of preimplantation blastocyts in the presence of LIF and BMP, mESCs fulfill all criteria of pluripotency. Therefore, they have been accepted as they are in "na" pluripotent state. On the other hand, EpiSCs are referred to as "primed" pluripotent cells because they exhibit only some pluripotency criteria.

Apart from the contribution to blastocyst chimeras, there are more differences between na and primed pluripotent states (Table 1): While na pluripotent cells show low susceptibility for primordial germ specification and high single-cell clonogenicity, primed cells display the reverse. While mESC colonies display compact dome-shaped morphologies, both hESCs and EpiSCs grow as a flat monolayer and the positive regulators of both the states are WNT and IGF. Na pluripotent cells require LIF/Stat3 and BMP4 signaling, whereas primed pluripotent cells need TGF-β, activin, FGF2, ERK1/2 signals. Interestingly enough, the two signals antagonize each other: BMP4 in EpiSCs and TGF-β, activin, FGF2, ERK1/2 in na cells induce differentiation. Na pluripotent cells generally do not express lineage specification markers (FGF5, Blimp1, Cer 1). However, these markers are positive in primed state cells with heterogeneous expression pattern. Moreover, primed pluripotent cells express MHC class I antigen, while na state cells do not. Na ESCs carry two active X chromosomes (XaXa) in female cells; in contrast primed EpiSCs have already undergone X chromosome inactivation (Nagy et al. 1990; Thomson et al. 1998; Ying et al. 2003; James et al. 2005; Brons et al. 2007; Tesar et al. 2007; Ying et al. 2008).

These studies proved that extrinsic stimuli are dispensable for the derivation, propagation and pluripotency of ESCs. They also showed that the cells have an innate program for self-replication. Ying et al. demonstrated that LIF and BMP could be dispensed with inducing inhibitors of particular signaling pathways. LIF and small molecule inhibitors of two protein kinases, ERK 1/2 and GSK3β (termed "2i") can replace serum by stimulating the WNT pathway. Therefore, 2i allows the maintenance of ESCs in fully defined medium without embryonic feeder cells. The results have been interpreted as the indication that the pluripotent and self-renewing state of ESCs is a "ground state", that is, a natural default state that need not be actively maintained (Ying et al. 2008).

Taken together, hESCs, EpiSCs and mESCs manifest themselves as potentially distinct cell types and hESCs may be most closely related to the post-implantation human epiblast (Wilmut et al. 2011). The generation of na hESCs will allow creating new opportunities for patient-specific researches. On the other hand, Guo et al. (2009) examined interconversion between mESCs and EpiSCs. They showed that when mESCs exposed to bFGF they could readily become EpiSCs. However,

EpiSCs do not change into ESCs with defined culture environment. Only forced expression of Klf4 in the presence of LIF and BMP could promote the conversion of EpiSCs into ESC-like cells (Guo et al. 2009). Another group, Bao and colleagues (2009) showed the reprogramming of advanced epiblast cells from embryonic day 5.5–7.5 mouse embryos to ESC-like cells in response to LIF-STAT3 signaling. The authors also reported that those reprogrammed ESCs could contribute to somatic tissues and germ cells in chimaeras unlike EpiSCs (Bao et al. 2009). Taken together, it is obvious that modulation of signaling by environmental changes are sufficient to interconvert these closely related cell types indicating that extrinsic growth factors could be dispensable for sustaining the pluripotent state (Ying et al. 2008). Moreover, Hanna et al. (2010a) converted ESCs into a more immature state with an active X-chromosome (XaXa) by ectopic induction of Oct4, Klf4 and Klf2, combined with LIF, GSK3β inhibitor and MEK inhibitor (Hanna et al. 2010a). These converted hESCs have similar growth properties, gene expression profiles and signaling pathway dependence with mESCs. Intriguingly enough, the recent establishment of preimplantation-derived EpiSCs cultured in human ESCs culture conditions supports this idea (Najm et al. 2011). It was also shown that hESCs with two active X chromosomes (XaXa) could be generated under hypoxic (5% oxygen) conditions (Lengner et al. 2010).

While the na and primed pluripotent states are inconvertible into each other and can be stabilized by appropriate culture conditions, these states have not been observed to coexist stably in the same culture conditions in both mouse and humans. However, it has been proposed that specific extrinsic and intrinsic factors can induce transitions between the states. Although the na state captured by Hanna et al. could be maintained only for limited passages, this study giving clear ideas about growth conditions should be improved in order to resume na ground state in genetically unmodified human cells (Hanna et al. 2010b).

Considerable research effort has been exerted to dissect the molecular functions of core pluripotency factors in the maintenance of pluripotency and establishment of a molecular association among pluripotent cell types. As already mentioned, since there has been important progress in dissecting culture-mediated signaling pathways; now the time is ripe to understand how signaling can be integrated to the transcriptional network.

2.2.1 *Transcriptional Network of Pluripotency*

ESCs have been investigated by very large-scale genomics and protein-DNA interaction studies to mechanistic studies of individual transcription factors. In addition to signaling requirements, particular transcription factors play major roles in establishing and maintaining pluripotency.

The most critical transcription factors of pluripotent state in ESCs are Oct4, Sox2 and Nanog and they are called the core transcription factors (Silva and Smith 2008). Transcription factors recognize specific DNA sequences and either activate or prevent transcription. They bind both to promoter-proximal DNA elements and to more distal regions (Young 2011).

Initial specification of pluripotent cells in vivo requires Oct4 expression (Nichols et al. 1998). While losing Oct4 expression leads to trophoectoderm differentiation, higher levels induce differentiation to mesoderm and endoderm (Niwa et al. 2000). Oct4 functions by forming heterodimer with Sox2 in ESCs. Sox2 binds to DNA sequences adjacent to the Oct4 binding sites (Avilion et al. 2003; Chambers and Tomlinson 2009). Sox2 is required for epiblast maintenance. Nanog, on the other hand, promotes a stable undifferentiated ESC state and it is needed for pluripotency to develop in ICM cells (Silva and Smith 2008). ESCs deficient in Nanog genes are more prone to differentiate but do not lose pluripotency per se. Nanog is essential for pluripotent cell specification during normal development and induction of pluripotency to finalize somatic cell reprogramming during induction of pluripotency (Theunissen and Silva 2011). Theunissen and Silva proposed that Nanog acts as a molecular switch to turn on the na pluripotent program in mammalian cells (Theunissen and Silva 2011).

Young has also suggested recently that there are two dominated concepts for our understanding of the core transcription factors in control of ESC state (Young 2011):

(1) The core transcription factors function together to positively regulate their own promoters, forming an interconnected autoregulatory loop.
(2) The core factors co-occupy and activate expression of genes necessary to maintain ESC state, while contributing to repression of genes encoding lineage-specific transcription factors whose absence helps prevent exit from the pluripotent state.

This interconnected autoregulatory loop could generate a bistable state for ESC: when the factors are expressed at appropriate levels positive-feedback-controlled gene expression program takes action. Alternatively, when any one of the core transcription factors is unavailable, functionally differentiation program is activated (Young 2011).

In recent years, there have been considerable efforts to understand pluripotency in a genome-wide manner as well as on a systems level to provide a global understanding of the ground state of ESC state and differentiation. Thus, the discovery of reprogramming of somatic mammalian cells into pluripotent state by overexpression of only four transcription factors had a tremendous effect in understanding the basic cell biology.

References

Avilion AA et al (2003) Multipotent cell lineages in early mouse development depend on SOX2 function. Genes Dev 17(1):126–140

Bao S et al (2009) Epigenetic reversion of post-implantation epiblast to pluripotent embryonic stem cells. Nature 461(7268):1292–1295

Brons IG et al (2007) Derivation of pluripotent epiblast stem cells from mammalian embryos. Nature 448(7150):191–195
Chambers I, Tomlinson SR (2009) The transcriptional foundation of pluripotency. Development 136(14):2311–2322
Chou YF et al (2008) The growth factor environment defines distinct pluripotent ground states in novel blastocyst-derived stem cells. Cell 135(3):449–461
Evans MJ, Kaufman MH (1981) Establishment in culture of pluripotential cells from mouse embryos. Nature 292(5819):154–156
Guan K et al (2006) Pluripotency of spermatogonial stem cells from adult mouse testis. Nature 440(7088):1199–1203
Guo G et al (2009) Klf4 reverts developmentally programmed restriction of ground state pluripotency. Development 136(7):1063–1069
Hanna J et al (2010a) Human embryonic stem cells with biological and epigenetic characteristics similar to those of mouse ESCs. Proc Natl Acad Sci U S A 107(20):9222–9227
Hanna J et al (2010b) Human embryonic stem cells with biological and epigenetic characteristics similar to those of mouse ESCs. Proc Natl Acad Sci U S A 107(20):9222–9227
James D et al (2005) TGFbeta/activin/nodal signaling is necessary for the maintenance of pluripotency in human embryonic stem cells. Development 132(6):1273–1282
Kanatsu-Shinohara M et al (2004) Generation of pluripotent stem cells from neonatal mouse testis. Cell 119(7):1001–1012
Lengner CJ et al (2010) Derivation of pre-X inactivation human embryonic stem cells under physiological oxygen concentrations. Cell 141(5):872–883
Matsui Y et al (1992) Derivation of pluripotential embryonic stem cells from murine primordial germ cells in culture. Cell 70(5):841–847
Meissner A et al (2008) Genome-scale DNA methylation maps of pluripotent and differentiated cells. Nature 454(7205):766–770
Nagy A et al (1990) Embryonic stem cells alone are able to support fetal development in the mouse. Development 110(3):815–821
Najm FJ et al (2011) Isolation of epiblast stem cells from preimplantation mouse embryos. Cell Stem Cell 8(3):318–325
Nichols J, Smith A (2009) Naive and primed pluripotent states. Cell Stem Cell 4(6):487–492
Nichols J et al (1998) Formation of pluripotent stem cells in the mammalian embryo depends on the POU transcription factor Oct4. Cell 95(3):379–391
Niwa H et al (2000) Quantitative expression of Oct-3/4 defines differentiation, dedifferentiation or self-renewal of ES cells. Nat Genet 24(4):372–376
Santos F et al (2002) Dynamic reprogramming of DNA methylation in the early mouse embryo. Dev Biol 241(1):172–182
Silva J, Smith A (2008) Capturing pluripotency. Cell 132(4):532–536
Surani MA (1999) Reprogramming a somatic nucleus by trans-modification activity in germ cells. Semin Cell Dev Biol 10(3):273–277
Tesar PJ et al (2007) New cell lines from mouse epiblast share defining features with human embryonic stem cells. Nature 448(7150):196–199
Theunissen TW, Silva JC (2011) Switching on pluripotency: a perspective on the biological requirement of Nanog. Philos Trans R Soc Lond B Biol Sci 366(1575):2222–2229
Thomson JA et al (1998) Embryonic stem cell lines derived from human blastocysts. Science 282(5391):1145–1147
Wilmut I et al (2011) The evolving biology of cell reprogramming. Philos Trans R Soc Lond B Biol Sci 366(1575):2183–2197
Ying QL et al (2003) BMP induction of Id proteins suppresses differentiation and sustains embryonic stem cell self-renewal in collaboration with STAT3. Cell 115(3):281–292
Ying QL et al (2008) The ground state of embryonic stem cell self-renewal. Nature 453(7194):519–523
Young RA (2011) Control of the embryonic stem cell state. Cell 144(6):940–954

Chapter 3
Induced Pluripotent Stem Cells (iPSCs)

3.1 Generation the First iPSCs

Takahashi and Yamanaka (2006) showed that overexpression of defined transcription factors can also convert cells (or nuclei) to pluripotent state. The idea was the factors that maintain pluripotency in ESCs might induce pluripotency in somatic cells. They tested 24 transcription factors on the basis of their important roles in mESCs. They introduced combinations of those genes into mouse fibroblasts through retroviral transduction. The cells were carrying an antibiotic-resistance gene that would only be expressed when F-box protein 15 gene (Fboxo15) was turned on. They designed this experiment so that this antibiotic-resistance gene could also be turned on if pluripotency was induced by the combination of 24 candidate genes. Surprisingly they found that only four of these factors were sufficient to generate ESC-like colonies. The combination of genes was octamer 3/4 (Oct4, also known as Pou5f1), SRY box-containing gene 2 (Sox2), Kruppel-like factor 4 (Klf4) and c-Myc (later called OKSM). These reprogrammed cells were called as induced pluripotent stem cells (iPSCs) (Takahashi and Yamanaka 2006).

Although the cells that were selected by their ability to express Fbxo15 had similarities to ESCs in morphology, growth properties, expression of important ESC marker genes such as SSEA-1 and Nanog and the ability to form teratomas in immunodeficient mice, they also differed in terms of global gene expression profiles and certain DNA methylation patterns. They failed to produce adult chimeric mice when injected into early mouse embryos. These characteristics indicated that these "first generation" iPSCs were not fully programmed (Yamanaka 2009a, b). Soon after, with the improved end points for the reprogramming process such as using Nanog or Oct4 as selector, instead of Fbxo15, they generated colonies with reactivated pluripotency genes, and these cells were also similar to ESCs and could contribute to adult chimeras (Okita et al. 2007). After one year, human fibroblasts were also reprogrammed with same transcription-factor genes (Takahashi et al. 2007; Wernig et al. 2007; Yu et al. 2007).

S. Yildirim, *Induced Pluripotent Stem Cells*, SpringerBriefs in Stem Cells

DOI: 10.1007/978-1-4614-2206-8_3,

Subsequently, several groups independently replicated the reprogramming of human fetal, neonatal and adult somatic cells into iPSCs (Aasen et al. 2008; Lowry et al. 2008; Park et al. 2008a, b). As an indication that transcriptional pathways of pluripotency have been well conserved during evolution, iPSCs have been generated from several species including human, rhesus monkey and murines (Takahashi et al. 2007; Liu et al. 2008; Li et al. 2009). More recently, iPSC lines were shown to generate "all iPSC" mice upon injection into tetraploid blastocysts (Boland et al. 2009; Kang et al. 2009; Zhao et al. 2009).

3.2 Reprogramming

3.2.1 Factor Delivery into Target Cells

iPSCs offer disease and patient-specific cells with the knowledge of the clinical history of the donor and they can be made with cells taken from all ages and conditions such as chronic disease in elder patients. The techniques used to generate iPSCs can affect the efficiency of reprogramming and the quality of resultant iPSCs. Reprogramming methods can be divided into several classes according to their mode of integration of exogenous genetic material into the host genome or those of non-integrative approaches (Stadtfeld and Hochedlinger 2010).

3.2.1.1 Integrating Approaches

Retroviruses

The delivery of Yamanaka factors (Oct4, Sox2, Klf4, c-Myc, OKSM) into mouse or human fibroblasts was originally achieved using Moloney murine leukemia virus (MMLV)-derived retroviruses. They stably integrated into the host cell genome (Takahashi and Yamanaka 2006; Okita et al. 2007; Wernig et al. 2007). The efficiency of iPSC generation using MMLV-derived retroviruses is $\sim$0.1% in mouse embryonic fibroblasts and $\sim$0.01% in human fibroblasts (Gonzalez et al. 2011). Silencing is important in order to obtain bona fide iPSCs. Only an iPSC that has upregulated the endogenous pluripotency gene network and downregulated the expression of transgenes can really be considered to be fully reprogrammed (Gonzalez et al. 2011). However, retrovirally derived iPSCs have numerous transgene integration sites in the host genome. Incomplete transgene silencing compromises the developmental potential of iPSCs (Brambrink et al. 2008). In addition, residual transgene expression or later reactivation of exogenously applied transcription factors could cause tumor formation at an alarming rate (Okita et al. 2007; Miura et al. 2009).

Lentiviruses

Having higher cloning capacities and infection efficiency than MMLV-based retroviruses, lentiviral delivery vectors have also been used to express reprogramming factors (Blelloch et al. 2007; Yu et al. 2007). As an advantage, they could infect both dividing and non-dividing cells. Although the efficiency of reprogramming is comparable to that of MMLV-derived retroviruses, lentiviruses are less effectively silenced in iPSCs that they generated (Yao et al. 2004). Constitutive lentiviral vectors could not solve the problem because they were even less efficiently repressed and caused a differentiation block (Blelloch et al. 2007; Yu et al. 2007).

Inducible lentivirus systems do not rely on direct factor delivery into target delivery. "Secondary" reprogramming systems use "primary" iPSC clones, generated with doxycycline (Dox)-inducible lentiviral vectors or transposons. Those primary iPSCs then differentiate into genetically homogenous somatic cells using either in vitro differentiation for human cells or blastocyst injection for mice (Maherali et al. 2008; Wernig et al. 2008). Since the primary cells are induced by a lentivirus-contained transcription factors expression cassette under the control of the Dox-inducible operator, "secondary" iPSCs formation are triggered when they are cultured in doxycycline-containing media (Maherali et al. 2008; Wernig et al. 2008). Although the efficiency of the system depends on the target cell type used, it is generally higher than that of primary infection (Stadtfeld and Hochedlinger 2010). Stadtfeld and Hochedlinger (2010) also reported that secondary systems have two main advantages: reprogramming of large quantities of genetically homogeneous cells, especially the one that is difficult to culture or transduce; and facilitating the comparison of genetically matched iPSCs derived from different somatic cells. However, there is still a disadvantage, which is heterogeneous expression patterns in secondary cell because of lentiviral transgenes that cause cumbersome screening of primary iPSC clones (Stadtfeld and Hochedlinger 2010). "Reprogrammable" mouse strains that may carry desired mutant backgrounds might have a striking effect for mechanistic studies of disease pathologies (Carey et al. 2010; Stadtfeld et al. 2010).

Single vector. Lentiviral or retroviral vectors designed to express polycistronic cassettes encoding all OKSM factors were more efficient than single factor expressing vectors in infecting different somatic cell types (Carey et al. 2009). Although this method minimizes the genomic integration, there is still tumorigenic risk after implantation. In particular c-Myc is a well-known oncogene and its reactivation could give rise to transgene-derived tumor formation in chimeric mice (Okita et al. 2007). Although removal of the c-Myc transgene from the reprogramming cocktail reduced the efficiency slightly, the chimeric mice produced with c-Myc-free iPSCs did not cause enhanced tumor formation (Nakagawa et al. 2008). However, the overexpression of Oct4 and Klf4 can still cause tumor formation and the retroviral insertion may disturb host gene structure and increase the tumor risk (Okita and Yamanaka 2010).

Standard DNA transfection techniques such as using liposomes or electroporation could be good alternatives to avoid the use of viral vectors. However, transduction efficiency is much lower than to that of viral transfections (Gonzalez et al. 2011).

3.2.1.2 Excisable Approaches

A virus-free method utilizing oriP/EBNA1 (Epstein–Barr nuclear antigen-1)-based episomal vectors or a piggyBac (PB)-based single vector reprogramming system has been used successfully for reprogramming human fibroblasts (Kaji et al. 2009; Yu et al. 2009). Another excisable method uses lentiviral vectors and Cre/*loxP* excisable linear transgenes that leave a genomic scar after Cre deletion. When polycistronic vectors were used, this approach produced the efficient generation of iPSCs from different cell types (Chang et al. 2009). Among integrative methodologies, although only PB system guarantees a precise deletion, it still needs verification of excised lines (Park et al. 2008b).

While the use of Cre-deletable or inducible lentiviruses has solved some problems, viral systems still lack the safety required for therapeutic applications.

3.2.1.3 Non-integrative Approaches

To minimize the risk of chromosomal disruptions, reprogramming protocols refined to eliminate genetic integration. For this purpose, three different approaches have been developed. First one utilizes the use of integrating vectors that can be subsequently removed from the genome as we have already discussed. Second one is using integration-free vectors and the last one is not to use nucleic acid-based vectors (Stadtfeld and Hochedlinger 2010). Unfortunately all of them are usually inefficient and poorly reproducible. (Gonzalez et al. 2011).

Among the methods that promote reprogramming without integration are adenoviral vectors that mediate transient expression of the exogenous reprogramming factors without genomic integration. One of the first attempts to generate integration-free iPSCs from adult mouse hepatocytes was reported by Stadtfeld et al. (2008), who used non-integrative adenoviral vectors. Using similar vectors Okita et al. (2008) and Zhou and Freed (2009) generated integration-free iPSCs lines from mouse embryonic fibroblasts and human fetal fibroblasts, respectively. Since there is no genomic integration with those systems, it is a proof of principle that transient expression of reprogramming factors is enough for generation of iPSCs from somatic cells (Stadtfeld and Hochedlinger 2010). Moreover firm support to conclusion came from recent experiments showing the absence of common integration sites in iPSCs produced with retroviruses or lentiviruses (Varas et al. 2009; Winkler et al. 2010). More recently iPSCs from human fibroblasts were successfully derived utilizing Sendai viral gene delivery systems

(Fusaki et al. 2009), as well as polycistronic minicircle vectors (Jia et al. 2010) and self-replicating selectable episomes (Yu et al. 2009).

DNA-Free Methods

In order to completely eliminate plasmid or viral vectors, more recently iPSCs have been generated by introducing four reprogramming factors in the form of purified recombinant proteins (Zhou et al. 2009; Kim et al. 2009a). However, its efficiency is extremely low, possibly due to low transduction efficiency and unstable expression (Okita and Yamanaka 2010). More recently Warren et al. (2010) showed more efficient reprogramming by using synthetic modified mRNA, which does not modify the genome. The iPSCs formed are called RNA-induced pluripotent cells (RiPSCs). Although the protocol is technically very complex, since this technology is RNA based, it completely eliminates the risk of genomic integration and insertional mutagenesis inherent to all DNA-based methodologies. This technology has the highest efficiency and kinetics to date (Walia et al. 2011).

Although the methods, which use non-integrating vectors, excisable genetic elements, small chemicals and/or proteins, leave no genetic footprint, the efficiency of iPSC generation with them ranges between 0.0001 and 0.0018%, which is approximately three orders of magnitude lower than those achieved by integrating vectors (0.1–1%) (Stadtfeld and Hochedlinger 2010). Therefore, reprogramming efficiencies have been trying to enhance with supplementing defined factors by additional genes or small chemicals (Masip et al. 2010).

3.2.1.4 Chemical Reprogramming

Small molecules are naturally occurring substances or entirely man-made compounds. They have been produced and screened for activity in diverse organisms, cell culture systems and molecular pathways. Their ease of use, specific, dose-dependent, rapid and reversible effects and precise temporal and functional control in vivo made them one of the most effective tools in biomedical research (Efe and Ding 2011). They have been using lately to facilitate somatic cell reprogramming. Efe and Ding (2011) explained how small molecules improve reprogramming: some has a global effect on cellular plasticity, whereas the others can modulate a specific signaling pathway to replace one or more reprogramming factors. On the other hand, there are also some compounds that work in both categories (Efe and Ding 2011). Omitting oncogenes (Klf4 and cMyc) (Nakagawa et al. 2008; Utikal et al. 2009; Kim et al. 2009b) or replacing one or two transcription factors with small chemicals such as VPA (a histone deacetylase inhibitor) (Huangfu et al. 2008) not only offer a safer clinical potential but also significantly enhance the efficiency of deriving iPS cells. More strikingly, Mali et al. (2010) have shown that low levels of butyrate (a HDAC inhibitor) improved reprogramming efficiency as much as 50-fold. Moreover, Lin et al. (2009) reported that combined treatment of fibroblasts with the Alk5 inhibitor SB431542, the MEK inhibitor PD0325901 and thiazovivin caused a great improvement (more than 200-fold) in OKSM reprogramming efficiency. (Lin et al. 2009)

3.2.1.5 MicroRNAs (miRNAs) in Reprogramming

miRNAs are 22 nt non-coding small RNAs that regulate expression of downstream targets by messenger RNA (mRNA) destabilization and translational inhibition (Subramanyam and Blelloch 2011). They are loaded into RNA-induced silencing complex (RISC) to exert a global gene-silencing function (Chu and Rana 2007). Multiple miRNAs are often found in clusters in the genome and expressed in a cell type-specific manner, similar to transcription factors. Since a single miRNA can target hundreds of mRNAs, the expression of single miRNA could regulate the cell fate decision powerfully (Sridharan and Plath 2011). Similar to small compounds, miRNAs are also shown that they are able to influence the efficiency of reprogramming. Anokye-Danso et al. (2011) have recently reported that lentivirally delivered miR-302 cluster, which includes miRs-302a-d and miR-367 alone can produce mouse and human iPSCs from fibroblasts. In addition Miyoshi et al. (2011) reprogrammed mouse and human adipose stromal cells by repeatedly transfecting a cocktail of seven miRNAs belonging the miR-302, miR-200 and miR-369 families. However, compared to the viral delivery of miRNA precursors in the study of Anokye and colleagues, the efficiency of reprogramming in latter research was considerably low. It has been suggested that in addition to their popular activities in reprogramming processes, screening cell type-specific miRNAs for transdifferentiation activities might also be very useful to modulate lineage conversion events (Sridharan and Plath 2011).

There are many alternative techniques for reprogramming. For example, Zhong et al. (2011) tested two apoptosis-inducing genes called "inducible suicide genes" or simply "suicide genes" to see if they are available for a potential method of safeguarding iPS cells. They evaluated the cell death-switch capacity of two suicide gene/drug systems, *YCD*/5-FC and *iCaspase*/AP20187, in non-human primate (*Macaca nemestrina*) iPS cells both in vitro and in vivo. The results of the study showed that a suicide gene/prodrug combination could safeguard iPSCs to overcome potential oncogenic transformation and/or persistence of pluripotent cells in iPSC-related cellular therapy (Zhong et al. 2011).

Yamanaka's team reported recently a non-integrative method using p53 suppression and non-transforming L-Myc with episomal plasmid vectors for reprogramming *HLA*-homozygous iPSCs. They used dental pulp cell lines from two putative human leukocyte antigen (HLA)-homozygous donors who match ~20% of the Japanese population at major HLA loci. As shown in this study dental pulp cells offer important safe sources for allografts, especially for cases in which transplants is likely to be needed soon after injury (Okita et al. 2011, Yildirim et al. 2011).

References

Aasen T et al (2008) Efficient and rapid generation of induced pluripotent stem cells from human keratinocytes. Nat Biotechnol 26(11):1276–1284

Anokye-Danso F et al (2011) Highly efficient miRNA-mediated reprogramming of mouse and human somatic cells to pluripotency. Cell Stem Cell 8(4):376–388

Blelloch R et al (2007) Generation of induced pluripotent stem cells in the absence of drug selection. Cell Stem Cell 1(3):245–247

Boland MJ et al (2009) Adult mice generated from induced pluripotent stem cells. Nature 461(7260):91–94

Brambrink T et al (2008) Sequential expression of pluripotency markers during direct reprogramming of mouse somatic cells. Cell Stem Cell 2(2):151–159

Carey BW et al (2009) Reprogramming of murine and human somatic cells using a single polycistronic vector. Proc Natl Acad Sci USA 106(1):157–162

Carey BW et al (2010) Single-gene transgenic mouse strains for reprogramming adult somatic cells. Nat Methods 7(1):56–59

Chang CW et al (2009) Polycistronic lentiviral vector for "hit and run" reprogramming of adult skin fibroblasts to induced pluripotent stem cells. Stem Cells 27(5):1042–1049

Chu CY, Rana TM (2007) Small RNAs: regulators and guardians of the genome. J Cell Physiol 213(2):412–419

Efe JA, Ding S (2011) The evolving biology of small molecules: controlling cell fate and identity. Philos Trans R Soc Lond B Biol Sci 366(1575):2208–2221

Fusaki N et al (2009) Efficient induction of transgene-free human pluripotent stem cells using a vector based on Sendai virus, an RNA virus that does not integrate into the host genome. Proc Jpn Acad Ser B Phys Biol Sci 85(8):348–362

Gonzalez F et al (2011) Methods for making induced pluripotent stem cells: reprogramming a la carte. Nat Rev Genet 12(4):231–242

Huangfu D et al (2008) Induction of pluripotent stem cells from primary human fibroblasts with only Oct4 and Sox2. Nat Biotechnol 26(11):1269–1275

Jia F et al (2010) A nonviral minicircle vector for deriving human iPS cells. Nat Methods 7(3):197–199

Kaji K et al (2009) Virus-free induction of pluripotency and subsequent excision of reprogramming factors. Nature 458(7239):771–775

Kang L et al (2009) iPS cells can support full-term development of tetraploid blastocyst-complemented embryos. Cell Stem Cell 5(2):135–138

Kim D et al (2009a) Generation of human induced pluripotent stem cells by direct delivery of reprogramming proteins. Cell Stem Cell 4(6):472–476

Kim JB et al (2009b) Direct reprogramming of human neural stem cells by OCT4. Nature 461(7264):649–653

Li W et al (2009) Generation of rat and human induced pluripotent stem cells by combining genetic reprogramming and chemical inhibitors. Cell Stem Cell 4(1):16–19

Lin T et al (2009) A chemical platform for improved induction of human iPSCs. Nat Methods 6(11):805–808

Liu H et al (2008) Generation of induced pluripotent stem cells from adult rhesus monkey fibroblasts. Cell Stem Cell 3(6):587–590

Lowry WE et al (2008) Generation of human induced pluripotent stem cells from dermal fibroblasts. Proc Natl Acad Sci USA 105(8):2883–2888

Maherali N et al (2008) A high-efficiency system for the generation and study of human induced pluripotent stem cells. Cell Stem Cell 3(3):340–345

Mali P et al (2010) Butyrate greatly enhances derivation of human induced pluripotent stem cells by promoting epigenetic remodeling and the expression of pluripotency-associated genes. Stem Cells 28(4):713–720

Masip M et al (2010) Reprogramming with defined factors: from induced pluripotency to induced transdifferentiation. Mol Hum Reprod 16(11):856–868
Miura K et al (2009) Variation in the safety of induced pluripotent stem cell lines. Nat Biotechnol 27(8):743–745
Miyoshi N et al (2011) Reprogramming of mouse and human cells to pluripotency using mature microRNAs. Cell Stem Cell 8(6):633–638
Nakagawa M et al (2008) Generation of induced pluripotent stem cells without Myc from mouse and human fibroblasts. Nat Biotechnol 26(1):101–106
Okita K, Yamanaka S (2010) Induction of pluripotency by defined factors. Exp Cell Res 316(16):2565–2570
Okita K et al (2007) Generation of germline-competent induced pluripotent stem cells. Nature 448(7151):313–317
Okita K et al (2008) Generation of mouse induced pluripotent stem cells without viral vectors. Science 322(5903):949–953
Okita K et al (2011) A more efficient method to generate integration-free human iPS cells. Nat Methods 8(5):409–412
Park IH et al (2008a) Reprogramming of human somatic cells to pluripotency with defined factors. Nature 451(7175):141–146
Park IH et al (2008b) Disease-specific induced pluripotent stem cells. Cell 134(5):877–886
Sridharan R, Plath K (2011) Small RNAs loom large during reprogramming. Cell Stem Cell 8(6):599–601
Stadtfeld M, Hochedlinger K (2010) Induced pluripotency: history, mechanisms, and applications. Genes Dev 24(20):2239–2263
Stadtfeld M et al (2008) Induced pluripotent stem cells generated without viral integration. Science 322(5903):945–949
Stadtfeld M et al (2010) A reprogrammable mouse strain from gene-targeted embryonic stem cells. Nat Methods 7(1):53–55
Subramanyam D, Blelloch R (2011) From microRNAs to targets: pathway discovery in cell fate transitions. Curr Opin Genet Dev 21(4):498–503
Takahashi K, Yamanaka S (2006) Induction of pluripotent stem cells from mouse embryonic and adult fibroblast cultures by defined factors. Cell 126(4):663–676
Takahashi K et al (2007) Induction of pluripotent stem cells from adult human fibroblasts by defined factors. Cell 131(5):861–872
Utikal J et al (2009) Sox2 is dispensable for the reprogramming of melanocytes and melanoma cells into induced pluripotent stem cells. J Cell Sci 122(Pt 19):3502–3510
Varas F et al (2009) Fibroblast-derived induced pluripotent stem cells show no common retroviral vector insertions. Stem Cells 27(2):300–306
Warren L et al (2010) Highly efficient reprogramming to pluripotency and directed differentiation of human cells with synthetic modified mRNA. Cell Stem Cell 7(5):618–630
Walia B et al (2011) Induced pluripotent stem cells: fundamentals and applications of the reprogramming process and its ramifications on regenerative medicine. Stem Cell Rev Jun 14. [Epub ahead of print]
Wernig M et al (2007) In vitro reprogramming of fibroblasts into a pluripotent ES-cell-like state. Nature 448(7151):318–324
Wernig M et al (2008) A drug-inducible transgenic system for direct reprogramming of multiple somatic cell types. Nat Biotechnol 26(8):916–924
Winkler T et al (2010) No evidence for clonal selection due to lentiviral integration sites in human induced pluripotent stem cells. Stem Cells 28(4):687–694
Yamanaka S (2009a) Elite and stochastic models for induced pluripotent stem cell generation. Nature 460(7251):49–52
Yamanaka S (2009b) A fresh look at iPS cells. Cell 137(1):13–17
Yao S et al (2004) Retrovirus silencing, variegation, extinction, and memory are controlled by a dynamic interplay of multiple epigenetic modifications. Mol Ther 10(1):27–36

Yildirim S et al (2011) Tooth regeneration: a revolution in stomatology and evolution in regenerative medicine. Int J Oral Sci 3(3):107–116

Yu J et al (2007) Induced pluripotent stem cell lines derived from human somatic cells. Science 318(5858):1917–1920

Yu J et al (2009) Human induced pluripotent stem cells free of vector and transgene sequences. Science 324(5928):797–801

Zhao XY et al (2009) iPS cells produce viable mice through tetraploid complementation. Nature 461(7260):86–90

Zhong B et al (2011) Safeguarding Nonhuman Primate iPS Cells With Suicide Genes. Mol Ther 19(9):1667–1675

Zhou W, Freed CR (2009) Adenoviral gene delivery can reprogram human fibroblasts to induced pluripotent stem cells. Stem Cells 27(11):2667–2674

Zhou H et al (2009) Generation of induced pluripotent stem cells using recombinant proteins. Cell Stem Cell 4(5):381–384

Chapter 4
Molecular Mechanisms of Pluripotency

4.1 Steps in Reprogramming

Combined with live imaging analysis, various tetracycline-inducible expression systems for the reprogramming factors are now allowing the probing to mechanistic insights of reprogramming. It is now being suggested that successful reprogramming requires stepwise transition through key intermediate steps (Papp and Plath 2011; Plath and Lowry 2011; Walia et al. 2011).

4.1.1 Increase in the Cell Cycle Rate

A high-resolution time-lapse imaging approach enabled to track that the first characterized event during reprogramming of fibroblasts is the increase of cell division rate (Smith et al. 2010). It indicated that certain cellular processes or events prime the cell before the exogenous transgenes reach to the genome to induce pluripotency (Jopling et al. 2011). Moreover, it has been shown that suppression of apoptosis and senescence is also helpful for successful reprogramming. Specifically, the silencing of p53 and p21 or CDKN2A have been observed upon reprogramming (Kawamura et al. 2009; Marion et al. 2009). Hence, apoptosis, senescence and cell cycle arrest are thought to be barriers for reprogramming (Papp and Plath 2011). However, the study used single cell analysis showed that although proliferation is induced in more cells than in the control sample, most of the p53-depleted cells undermine the reprogramming path later. And the resultant efficiency of the reprogramming was very low when normalized to the number of cells that initially responded (Smith et al. 2010).

The considerable variation between cells in response to induced reprogramming can be shown with the help of a cell sorting experiment for Thy1 (a marker for fibroblast phenotype). In the study by Stadtfeld et al. (2008b) it has been shown that, although almost half of the cell had lost Thy1 expression after five days of

S. Yildirim, *Induced Pluripotent Stem Cells*, SpringerBriefs in Stem Cells
DOI: 10.1007/978-1-4614-2206-8_4,

treatment, only very small fraction of those cells become fully reprogrammed at the end of the process. However, Hanna et al. (2009) displayed that extended induction of transgene expression could induce most, if not all, cells to iPSCs.

4.1.2 Morphological Changes

Rapid cycling cells get smaller over time and continue to grow as monolayers although most cells expressing the reprogramming factors fail to successfully induce the first morphological change of proper reprogramming events, remain fibroblast like and often undergo apoptosis, senescence and cell cycle arrest (Plath and Lowry 2011).

ESCs and their in vivo counterparts, the epiblast progenitor cells of the pre-implantation blastocyst, are epithelial in nature, therefore they have close cell–cell contacts, high proliferation rate with an extremely short G1 cell cycle phase, and large nucleus to cytoplasm ratio (Papp and Plath 2011). At the initial phases of reprogramming, fast cycling cells cluster tightly and undergo changes that correspond to loss of mesenchymal features and acquisition of epithelial cell characteristics (Li et al. 2010). This is exactly the opposite of the epithelial-mesenchymal transitioning (EMT) of developmental pluripotency to differentiation transition, during reprogramming (Li et al. 2010; Samavarchi-Tehrani et al. 2010). Therefore, signaling pathways known to promote mesenchymal-epithelial transition (MET) should increase the efficiency of reprogramming. Accordingly, it has been shown that inhibition of TGF-β improves reprogramming, because TGF-β activity prevents MET (Li et al. 2010). In addition it has been shown that BMP signaling enhances reprogramming through the up-regulation of pro-MET microRNAs (Samavarchi-Tehrani et al. 2010). Moreover, E-cadherin-mediated cell–cell contacts are required and its knockdown interferes with reprogramming (Chen et al. 2010).

4.1.3 Late Events Toward Pluripotency

The genes that are related to epithelial cell character are activated for pluripotency. While alkaline phosphatase and SSEA-1 are considered as intermediate stage markers, Nanog in the mouse and TRA-1-60 in human cells are final stage markers for pluripotency (Okita et al. 2007; Wernig et al. 2007; Mikkelsen et al. 2008; Stadtfeld et al. 2008b; Chan et al. 2009). It has been postulated that expression of reprogramming factors initiate sequential and fairly synchronous order. Although iPSC state needs continuous expression of the reprogramming factors to overcome roadblocks toward pluripotency, maintenance of iPSCs is independent of factor overexpression (Brambrink et al. 2008; Stadtfeld et al. 2008b; Papp and Plath 2011; Plath and Lowry 2011).

4.2 Mechanisms in Reprogramming

In the process of reprogramming it has been shown that a genome-wide alteration of epigenetic marks, regulation of cell cycle checkpoints and MET are required for the acquisition of pluripotency (Samavarchi-Tehrani et al. 2010; Plath and Lowry 2011). While reprogramming factors reset the cellular phenotype from the inside, it also requires extrinsic signals provided by the ESC culture conditions including growth factors, cytokines and other signals provided by the cell culture medium, fetal bovine serum and feeder cells. How extrinsic signals are integrated with intrinsically acting factors is not entirely clear, and it is an active area of investigation (Ralston and Rossant 2010).

4.2.1 Genetic Factors

The reprogramming of somatic cells required using of genetic factors, namely Oct4, Nanog, Sox2, Klf4, c-Myc and Lin28 (Takahashi and Yamanaka 2006; Yu et al. 2007). It has been shown previously that Nanog, Lin28 and c-Myc can be dispensable for reprogramming (Yu et al. 2007; Nakagawa et al. 2008; Wernig et al. 2008b).

Theunissen and Silva (2011) reported that "Nanog counteracts differentiation but is ultimately dispensable for maintenance of pluripotency in ESCs". Accordingly, although Nanog was not among the core transcription factors of reprogramming, it is essential to finalize reprogramming. The authors proposed that Nanog acts as a molecular switch to turn on the naïve pluripotent programme in mammalian cells (Theunissen and Silva 2011). Furthermore, it has been shown that Nanog overexpression could accelerate reprogramming in a predominantly cell division rate independent manner (Silva et al. 2006).

Myc, unlike Oct4, Sox2 and Klf4, is not involved in the upregulation of the pluripotency network during the final step of reprogramming. The fact that Myc might enhance but is not absolutely required for the transciption of pluripotency target genes, explains why this factor is dispensable for the reprogramming but still is able to enhance the efficiency and the kinetics of the process (Nakagawa et al. 2008; Wernig et al. 2008b; Plath and Lowry 2011).

Appropriate chemical compounds in culture media may be substituted Sox2 and Klf4. These two factors were also dispensable, when the somatic cells, which have high expression of Sox2 and Klf4 are used for reprogramming (Aoi et al. 2008; Huangfu et al. 2008; Li et al. 2009).

The last element of core transcriptional regulatory circuitry, Oct4, has been shown to be the most important factor in the reprogramming cocktail. It is the only one of the initial four factors, which has been shown to be essential for reprogramming (Walia et al. 2011). However, a recent study has showed that NR5A2, also known as LRH-1 (liver receptor homolog-1), can replace Oct4 with improved reprogramming efficiency (Heng et al. 2010).

It is obvious that the effects of main reprogramming factors are regulated and mediated by other receptors, transcription factors, growth factors, enzymes, other proteins and chemicals (Walia et al. 2011).

4.2.2 Signaling Pathways

4.2.2.1 Wnt Signaling

Some of the components of this evolutionary conserved pathway have been shown to enhance reprogramming activity. A downstream effector of the Wnt pathway, an inhibitor of glycogen synthase kinase-3 (GSK3) called 6-bromoindirubin-3′-oxime or BIO, has been shown to maintain pluripotency in human and mouse ESCs (Sato et al. 2004). Addition of Wnt3a to culture media enhance reprogramming efficiency (Marson et al. 2008). Another GSK3 inhibitor, CHIR99021 compensates Sox2 absence in hiPSC programming (Li et al. 2009). Giving a big credit to Wnt/β-catenin signaling pathway mediated activation of transcription, Abu-Remaileh et al. (2010) showed that Oct4 interacts with β-catenin and modulates Wnt signaling to maintain stem cell identity and regulate cell fate decision.

4.2.2.2 TGF-β/Activin/Nodal Signaling

As one of the most studied pathway, TGF-β/Activin/Nodal plays important roles in the maintenance of pluripotent state. FGF and TGFβ/Activin signaling promote self-renewal of hESCs. Nanog has been shown as a critical molecule for pluripotency and a direct target of TGF-β/Activin pathway (Xu et al. 2008). By the binding of Smad coactivators to the Nanog proximal promoter, Nanog expression is upregulated (Theunissen and Silva 2011). While inhibition of TGF-β signaling using a TGF-β receptor/Alk5 kinase inhibitor enhances the efficiency of mouse iPSC reprogramming and this inhibitor could replace with Sox2 and c-Myc. However, similar effect was not seen in hiPSCs (Maherali and Hochedlinger 2009). It has been shown that TGF-β signaling inhibition leads mesenchymal transition in hESCs (James et al. 2005).

4.2.2.3 BMP Signaling

BMPs belong to the TGF-β superfamily. They are involved in the regulation of cell proliferation, differentiation and apoptosis and therefore play essential roles during embryonic development and pattern formation (Massague 1998). Xu et al. (2008) showed that both TGFβ- and BMP-responsive SMADs can bind with the Nanog proximal promoter causing an enhancement to Nanog promoter activity and this activity is decreased by BMP signaling. Differentiation of hESCs is promoted by BMP signaling (Xu et al. 2008). Another BMP family member is growth and differentiation factor 3 or GDF3 and it has been shown to block BMP signaling in both hESCs and mESCs (Levine and Brivanlou 2006).

4.2.2.4 p53 Pathway

Several groups reported that p53, one of the well-studied tumor suppressor transcription factors, acts as a barrier to somatic cell reprogramming (Kawamura et al. 2009; Li et al. 2009; Marion et al. 2009; Utikal et al. 2009). Hong et al. (2009) proposed that the p53-p21 pathway serves as a safeguard not only in tumorigenicity, but also in iPS cell generation (Hong et al. 2009). On the other hand, Hanna et al. (2009) showed that most cells are capable of becoming iPSCs without depleting p53 or immortalizing the cells (Hanna et al. 2009). Moreover, monitoring the effect of p53 knockdown at the single cell level showed that most of the p53-depleted cells derail from reprogramming path and lower overall reprogramming efficiency (Smith et al. 2010). It has been suggested that future efforts for more successful reprogramming should select target cells with lower levels of p53 and/or higher proliferative ability, rather than silencing the p53 pathway (Tapia and Scholer 2010). Accordingly, Yamanaka's team has already published a simple and efficient method, using p53 suppression and nontransforming c-Myc to reprogram human somatic cells (Okita et al. 2011).

4.3 Dynamics of Direct Reprogramming

In general, induced pluripotent somatic cell reprogramming is an inefficient process even when different strategies are used. There is a latency of approximately 5–10 days before the first iPSCs appears. This delay and the low efficiency are indicators of stochastic mechanisms involved in inducing reprogramming (Hanna et al. 2009). While low efficiency is a strong barrier for applying human iPSCs to medicine, the stochastic and rate-limiting epigenetic mechanisms show that with a balanced reprogramming factor stoichiometry, nearly every cell is able to generate iPSCs (Hanna et al. 2010b).

Yamanaka (2009a, b) proposed recently that only a few cells in a primary cell culture might be competent for reprogramming. In fact, the retroviral vector can infect over 90% of fibroblasts, however only a small number of iPSC colonies emerged with the efficiency of 0.001%. This low efficiency suggests that the origin of iPSCs may be emerging from very small number of cells from the tissue. The candidates for these cells are adult stem or progenitor cells, as they are rare and developmentally closer to pluripotent cells than differentiated cells. It was also thought that for successful reprogramming, activation of additional genes by insertional mutagenesis might be crucial (Yamanaka 2009a, b). However, this model (called elite model) could not sustain longer, since human and mouse iPSCs generated from several terminally differentiated somatic cells, including B and T lymphocytes (Hanna et al. 2008; Eminli et al. 2009) as well as pancreatic β cells (Stadtfeld et al. 2008a). Furthermore Okita et al. (2008) and Stadtfeld et al. (2008a) showed simultaneously that insertional mutagenesis is not an essential step of reprogramming process.

Therefore Yamanaka suggests another theory in stochastic model. In this model he offered most, if not all, differentiated cells have the potential to become iPSCs by four factor introduction (Yamanaka 2009a, b). Likewise, Meissner et al. (2007) showed previously that even morphologically similar iPSC colonies start express Oct4 at different times. These data clearly show reprogramming events are stochastic in nature (Meissner et al. 2007). Besides, in contrast to claim that the differentiation state itself may also influence the susceptibility of cells to form iPSCs (Eminli et al. 2009), there are also reports that demonstrate that iPSCs do not prefer less differentiated cells and controlling in vitro cell plating efficiency, growth expansion and gene delivery methods are critical for this purpose (Hanna et al. 2009).

Although the exact mechanisms are still unknown, reprogramming is a very complex sequence of events requires silencing of the somatic cell program than resetting of the pluripotency programs. Apart from the genetic factors and signaling pathways, epigenetic modulations and regulation clearly play role for breaking the epigenetic barriers then reaching self-renewal and pluripotency.

4.4 Epigenetic Modifications

Epigenetics describes mitotically heritable modifications of DNA or chromatin without altering the nucleotide sequence. DNA methylation, histone modification, histone variants and non-coding RNAs are mechanisms of epigenetic regulation (Bird 2002). A central question in the field of ESCs pluripotency is the extent to which epigenetic marks regulate, rather than simply reflect, the pluripotent state in vitro (Meissner 2010). It has been suggested that ESCs have unique signatures including highly expressed DNA methyltransferases (Dnmt1, 2, 3a, 3b and 3 l) and the core PRC1 and 2 subunits, distinct CpG distribution than somatic cells and ICM and finally different distribution and enrichment of various histone modifications (Meissner 2010). However, their individual regulatory roles are far from our intuitive understanding. Despite significant advances in technology, it is still difficult to describe lineage specification and the associated global epigenetic remodeling for many cell types in vitro. Therefore, it is still fascinating how a somatic cell changes its epigenetic state to pluripotency by repealing the existing epigenetic state without passing through normal development or complete resetting of all marks (Meissner 2010). It has been suggested that somatic cell reprogramming involves massive reconfiguration of chromatin structure, from DNA methylation to histone modifications to nucleosome remodeling (Ang et al. 2011).

Candidates for the epigenetic modifications in de-differentiation process from somatic cells to iPSCs are promoter CpG DNA methylation, and alteration in histone modifications in relation with this importance of PcG mediated repression of genes. Genome-wide approaches show that the pattern of H3K9me3 within promoter regions is different between hESC and hiPSCs (Maherali et al. 2007; Fouse et al. 2008; Doi et al. 2009). The comparison of iPSCs and ESCs has revealed that the conversion of monovalent histone methylation marks such as H3K4me3 or H3K27me3 to bivalent marks in the developmental genes (Bernstein et al. 2006). The loss of DNA methylation along with the reactivation of transcription of pluripotency genes is another important steps in the acquisition of pluripotency. For example, inhibition of DNA methylation with the DNMT inhibitor 5-azacytidine; inhibition of histone deacetylation using the HDAC inhibitors VPA and trichostatin A improve reprogramming efficiency. Moreover, using a K4 demethylase LSD1 inhibitor Parnate (tranylcyprome) and the induction of histone H3K9 hypomethylation using the G9a methyltransferase chemical inhibitor BIX-01294 were successful to increase reprogramming efficiency (Huangfu et al. 2008; Mikkelsen et al. 2008; Shi et al. 2008). Besides, it is logical to think that downregulation of epigenetic factors causes removal of epigenetic barriers preventing reprogramming. Likewise, loss of CHD1 in mouse somatic cells and a transcriptional regulator PRDM14 in human cells results in decreased reprogramming efficiency (Chia et al. 2010).

Two groups have recently showed that the DNA methylation pattern of the original cell persisted in mouse iPSCs and affected their differentiation potential; iPSCs from blood more easily differentiated toward blood lineages than iPSCs derived from fibroblasts (Kim et al. 2010; Polo et al. 2010). Although continued passaging of iPSCs could erase this epigenetic memory (Polo et al. 2010), residual DNA methylation within lineage-specific genes of iPSCs demonstrates that resetting this mark is fundamental to reprogramming (Plath and Lowry 2011). Furthermore, Lister et al. (2011) showed that reprogramming also induced aberrant methylation that seems to be specific to iPSC state.

4.5 Similarities and Differences Between iPSCs and ESCs

The characteristics of somatic cells are limited proliferation, methylation of pluripotency genes, displaying tissue-specific cell morphology, inactivation of X chromosome, active G1 cell cycle checkpoint and expression of somatic cell specific markers. On the other hand, pluripotent cells robustly show self-renewal, ESC morphology, and reactivation of pluripotency genes by demethylation, X-chromosome reactivation in female cells, telomerase activity and finally loss of G1 checkpoint (Cox and Rizzino 2010). Although a number of studies have clearly showed that iPSCs are highly similar to ESCs (Mikkelsen et al. 2008; Okita et al. 2008; Soldner et al. 2009), extensive examination of chromatin state of both has shown that there are consistent and functionally relevant differences (Plath and

Lowry 2011). It has been demonstrated that gene expression signatures of virally programmed iPSCs are distinguished from that of ESCs (Lowry et al. 2008; Maherali et al. 2008; Chin et al. 2009; Soldner et al. 2009). However, human iPSCs generated without viral vectors or genomic insertions still displayed retained potential transcriptional signatures (Marchetto et al. 2010). Wang et al. (2010) compared transcriptomes of fibroblasts, partially reprogrammed iPSCs, ESCs and iPSCs using microarray data. They concluded that most of the reprogrammed iPSC lines were similar to ESC lines. Especially, the transcriptomes of the iPSCs derived by episomal-based non-integrating plasmids (Yu et al. 2009) were much closer to that of ESCs, unlike that of retroviral-derived iPSCs (Lowry et al. 2008; Maherali et al. 2008; Chin et al. 2009; Soldner et al. 2009). Moreover, the extent of overlapping implemented in the microarray platform was determined by showing that ESCs and iPSCs express 2390 common genes, with only 249 and 684 genes expressed exclusively in those cells respectively. In general, iPSCs share more genes in common with the fibroblasts from which they were derived (Wang et al. 2010).

Given that the reprogramming process is expected to remove any epigenetic alterations associated with originated cells from nuclear transfer (NT) and ESCs, which have reported faithfully erased any epigenetic marks present in donor cells (Okita et al. 2007; Maherali et al. 2008; Mikkelsen et al. 2008), the data showing iPSCs share more common genes with their parental fibroblasts was quite surprising. In fact, in 19 July 2010 simultaneously published two papers proved that even rigorously selected early-passage iPSCs could retain epigenetic marks that are characteristic of the donor cell (Kim et al. 2010; Polo et al. 2010). Plath and Lowry (2011) stated that epigenetic memory might be a form of incomplete reprogramming.

Furthermore, Urbach et al. compared hESC lines isolated from an embryo affected by Fragile X syndrome and reprogrammed fibroblasts of affected individuals. They observed that the *FMR1* gene remained silenced in the iPSCs. While those iPSCs showed stringent pluripotency criteria, the *FMR1* locus was obviously resistant to reprogramming (Urbach et al. 2010). This study shows that human ES and iPS cells are not identical, even in some circumstances ESCs may more faithfully reflect the disease process (Unternaehrer and Daley 2011). Moreover, Bock et al. (2011) has used quantification of molecular similarity by a "scorecard" included DNA methylome, transcriptome and differentiation studies and reported that although ESCs exhibit significant variability across individual lines, iPSCs are more variable at the molecular level than ESCs (Lister et al. 2011). Plath and Lowry (2011) interpreted those differences as follows:

> These methylation differences between ESCs and iPSCs are associated with differences at the transcriptional level that can be found after many passages and might affect the differentiation behaviour of cells.

Utilizing both "forward" and "reverse" genetic approaches with the aid of iPSCs offer exciting prospects for dissecting molecular mechanisms of commitment and differentiation in a cell lineage and will help to understand rewiring the regulatory networks that are active in somatic and pluripotent cells (Hemberger et al. 2009; Huang 2009; Saha and Jaenisch 2009). Future work should focus on finding out when and how the lineage-specific genetic and epigenetic marks arise.

References

Abu-Remaileh M et al (2010) Oct-3/4 regulates stem cell identity and cell fate decisions by modulating Wnt/beta-catenin signalling. EMBO J 29(19):3236–3248
Ang YS et al (2011) Stem cells and reprogramming: breaking the epigenetic barrier? Trends Pharmacol Sci 32(7):394–401
Aoi T et al (2008) Generation of pluripotent stem cells from adult mouse liver and stomach cells. Science 321(5889):699–702
Bernstein BE et al (2006) A bivalent chromatin structure marks key developmental genes in embryonic stem cells. Cell 125(2):315–326
Bird A (2002) DNA methylation patterns and epigenetic memory. Genes Dev 16(1):6–21
Bock C et al (2011) Reference maps of human ES and iPS cell variation enable high-throughput characterization of pluripotent cell lines. Cell 144(3):439–452
Brambrink T et al (2008) Sequential expression of pluripotency markers during direct reprogramming of mouse somatic cells. Cell Stem Cell 2(2):151–159
Chan EM et al (2009) Live cell imaging distinguishes bona fide human iPS cells from partially reprogrammed cells. Nat Biotechnol 27(11):1033–1037
Chen T et al (2010) E-cadherin-mediated cell–cell contact is critical for induced pluripotent stem cell generation. Stem Cells 28(8):1315–1325
Chia N et al. (2010) A genome-wide RNAi screen reveals determinants of human embryonic stem cell identity. Nature 468(7321):316–320
Chin MH et al (2009) Induced pluripotent stem cells and embryonic stem cells are distinguished by gene expression signatures. Cell Stem Cell 5(1):111–123
Cox JL, Rizzino A (2010) Induced pluripotent stem cells: what lies beyond the paradigm shift. Exp Biol Med (Maywood) 235(2):148–158
Doi A et al (2009) Differential methylation of tissue- and cancer-specific CpG island shores distinguishes human induced pluripotent stem cells, embryonic stem cells and fibroblasts. Nat Genet 41(12):1350–1353
Eminli S et al (2009) Differentiation stage determines potential of hematopoietic cells for reprogramming into induced pluripotent stem cells. Nat Genet 41(9):968–976
Fouse SD et al (2008) Promoter CpG methylation contributes to ES cell gene regulation in parallel with Oct4/Nanog, PcG complex, and histone H3 K4/K27 trimethylation. Cell Stem Cell 2(2):160–169
Hanna J et al (2008) Direct reprogramming of terminally differentiated mature B lymphocytes to pluripotency. Cell 133(2):250–284
Hanna J et al (2009) Direct cell reprogramming is a stochastic process amenable to acceleration. Nature 462(7273):595–601
Hanna JH et al (2010) Pluripotency and cellular reprogramming: facts, hypotheses, unresolved issues. Cell 143(4):508–525
Hemberger M et al (2009) Epigenetic dynamics of stem cells and cell lineage commitment: digging Waddington's canal. Nat Rev Mol Cell Biol 10(8):526–537
Heng JC et al (2010) The nuclear receptor Nr5a2 can replace Oct4 in the reprogramming of murine somatic cells to pluripotent cells. Cell Stem Cell 6(2):167–174
Hong H et al (2009) Suppression of induced pluripotent stem cell generation by the p53-p21 pathway. Nature 460(7259):1132–1135
Huang S (2009) Reprogramming cell fates: reconciling rarity with robustness. Bioessays 31(5):546–560

Huangfu D et al (2008) Induction of pluripotent stem cells from primary human fibroblasts with only Oct4 and Sox2. Nat Biotechnol 26(11):1269–1275
James D et al (2005) TGFbeta/activin/nodal signaling is necessary for the maintenance of pluripotency in human embryonic stem cells. Development 132(6):1273–1282
Jopling C et al (2011) Dedifferentiation, transdifferentiation and reprogramming: three routes to regeneration. Nat Rev Mol Cell Biol 12(2):79–89
Kawamura T et al (2009) Linking the p53 tumour suppressor pathway to somatic cell reprogramming. Nature 460(7259):1140–1144
Kim K et al (2010) Epigenetic memory in induced pluripotent stem cells. Nature 467(7313):285–290
Levine AJ, Brivanlou AH (2006) GDF3, a BMP inhibitor, regulates cell fate in stem cells and early embryos. Development 133(2):209–216
Li W et al (2009) Generation of rat and human induced pluripotent stem cells by combining genetic reprogramming and chemical inhibitors. Cell Stem Cell 4(1):16–19
Li R et al (2010) A mesenchymal-to-epithelial transition initiates and is required for the nuclear reprogramming of mouse fibroblasts. Cell Stem Cell 7(1):51–63
Lister R et al (2011) Hotspots of aberrant epigenomic reprogramming in human induced pluripotent stem cells. Nature 471(7336):68–73
Lowry WE et al (2008) Generation of human induced pluripotent stem cells from dermal fibroblasts. Proc Natl Acad Sci USA 105(8):2883–2888
Maherali N et al (2007) Directly reprogrammed fibroblasts show global epigenetic remodeling and widespread tissue contribution. Cell Stem Cell 1(1): 55–70
Maherali N, Hochedlinger K (2009) Tgfbeta signal inhibition cooperates in the induction of iPSCs and replaces Sox2 and cMyc. Curr Biol 19(20):1718–1723
Maherali N et al (2008) A high-efficiency system for the generation and study of human induced pluripotent stem cells. Cell Stem Cell 3(3):340–345
Marchetto MC et al (2010) A model for neural development and treatment of Rett syndrome using human induced pluripotent stem cells. Cell 143(4):527–539
Marion RM et al (2009) A p53-mediated DNA damage response limits reprogramming to ensure iPS cell genomic integrity. Nature 460(7259):1149–1153
Marson A et al (2008) Wnt signaling promotes reprogramming of somatic cells to pluripotency. Cell Stem Cell 3(2):132–135
Massague J (1998) TGF-beta signal transduction. Annu Rev Biochem 67:753–791
Meissner A et al (2007) Direct reprogramming of genetically unmodified fibroblasts into pluripotent stem cells. Nat Biotechnol 25(10):1177–1181
Meissner A (2010) Epigenetic modifications in pluripotent and differentiated cells. Nat Biotechnol 28(10):1079–1088
Mikkelsen TS et al (2008) Dissecting direct reprogramming through integrative genomic analysis. Nature 454(7200):49–55
Nakagawa M et al (2008) Generation of induced pluripotent stem cells without Myc from mouse and human fibroblasts. Nat Biotechnol 26(1):101–106
Okita K et al (2007) Generation of germline-competent induced pluripotent stem cells. Nature 448(7151):313–317
Okita K et al (2008) Generation of mouse induced pluripotent stem cells without viral vectors. Science 322(5903):949–953
Okita K et al (2011) A more efficient method to generate integration-free human iPS cells. Nat Methods 8(5):409–412
Papp B, Plath K (2011) Reprogramming to pluripotency: stepwise resetting of the epigenetic landscape. Cell Res 21(3):486–501
Plath K, Lowry WE (2011) Progress in understanding reprogramming to the induced pluripotent state. Nat Rev Genet 12(4):253–265
Polo JM et al (2010) Cell type of origin influences the molecular and functional properties of mouse induced pluripotent stem cells. Nat Biotechnol 28(8):848–855
Ralston A, Rossant J (2010) The genetics of induced pluripotency. Reproduction 139(1):35–44

Saha K, Jaenisch R (2009) Technical challenges in using human induced pluripotent stem cells to model disease. Cell Stem Cell 5(6):584–595

Samavarchi-Tehrani P et al (2010) Functional genomics reveals a BMP-driven mesenchymal-to-epithelial transition in the initiation of somatic cell reprogramming. Cell Stem Cell 7(1):64–77

Sato N et al (2004) Maintenance of pluripotency in human and mouse embryonic stem cells through activation of Wnt signaling by a pharmacological GSK-3-specific inhibitor. Nat Med 10(1):55–63

Shi Y et al (2008) A combined chemical and genetic approach for the generation of induced pluripotent stem cells. Cell Stem Cell 2(6):525–528

Silva J et al (2006) Nanog promotes transfer of pluripotency after cell fusion. Nature 441(7096):997–1001

Smith ZD et al (2010) Dynamic single-cell imaging of direct reprogramming reveals an early specifying event. Nat Biotechnol 28(5):521–526

Soldner F et al (2009) Parkinson's disease patient-derived induced pluripotent stem cells free of viral reprogramming factors. Cell 136(5):964–977

Stadtfeld M et al (2008a) Induced pluripotent stem cells generated without viral integration. Science 322(5903):945–949

Stadtfeld M et al (2008b) Defining molecular cornerstones during fibroblast to iPS cell reprogramming in mouse. Cell Stem Cell 2(3):230–240

Takahashi K, Yamanaka S (2006) Induction of pluripotent stem cells from mouse embryonic and adult fibroblast cultures by defined factors. Cell 126(4):663–676

Tapia N, Scholer HR (2010) p53 connects tumorigenesis and reprogramming to pluripotency. J Exp Med 207(10):2045–2048

Theunissen TW, Silva JC (2011) Switching on pluripotency: a perspective on the biological requirement of Nanog. Philos Trans R Soc Lond B Biol Sci 366(1575):2222–2229

Unternaehrer JJ, Daley GQ (2011) Induced pluripotent stem cells for modelling human diseases. Philos Trans R Soc Lond B Biol Sci 366(1575):2274–2285

Urbach A et al (2010) Differential modeling of fragile X syndrome by human embryonic stem cells and induced pluripotent stem cells. Cell Stem Cell 6(5):407–411

Utikal J et al (2009) Sox2 is dispensable for the reprogramming of melanocytes and melanoma cells into induced pluripotent stem cells. J Cell Sci 122(Pt 19):3502–3510

Walia B et al (2011) Induced pluripotent stem cells: fundamentals and applications of the reprogramming process and its ramifications on regenerative medicine. Stem Cell Rev Jun 14. [Epub ahead of print]

Wang Y et al (2010) A transcriptional roadmap to the induction of pluripotency in somatic cells. Stem Cell Rev 6(2):282–296

Wernig M et al (2007) In vitro reprogramming of fibroblasts into a pluripotent ES-cell-like state. Nature 448(7151):318–324

Wernig M et al (2008) c-Myc is dispensable for direct reprogramming of mouse fibroblasts. Cell Stem Cell 2(1):10–12

Xu RH et al (2008) NANOG is a direct target of TGFbeta/activin-mediated SMAD signaling in human ESCs. Cell Stem Cell 3(2):196–206

Yamanaka S (2009a) Elite and stochastic models for induced pluripotent stem cell generation. Nature 460(7251):49–52

Yamanaka S (2009b) A fresh look at iPS cells. Cell 137(1):13–17

Yu J et al (2007) Induced pluripotent stem cell lines derived from human somatic cells. Science 318(5858):1917–1920

Yu J et al (2009) Human induced pluripotent stem cells free of vector and transgene sequences. Science 324(5928):797–801

Chapter 5
Modeling Disease in a Dish

Main idea for therapeutic approaches by iPSCs at the beginning was the fact that patient-specific iPSCs provide important information for inherited human disorders because pluripotent stem cells are capable of differentiation into most, if not all cell types. This idea was deeply relied on the studies of directed differentiation of subtypes and genetically defined ESCs from animal models (Gearhart 1998). Moreover, human ESC biology has been pursuing generating mutant human ESC lines as disease models since Thomson et al. (1998) derived human ESC lines in 1998. With the known disease-associated genetic loci and explicit disease phenotype, genetically modified human ESCs could help to cell replacement therapies and modeling human diseases (Saha and Jaenisch 2009).

However, the use of ESCs has several limitations not only in political, religious, ethical and moral concerns on the destruction of human embryos but also inefficient methods to generate genetically modified human ESCs. For example, human ESC generation via PGD embryos is available for only a limited number of diseases and lack of proper techniques are still challenging. Besides, only a few monogenic diseases could be detectable via PGD and there is no consistency between severity and clinical symptoms of those diseases from patient to patient due to variable penetrance (Colman and Dreesen 2009). Other alternatives such as generating individual pluripotent stem cells by NT, cell fusion with ESCs and treatment with extracts of pluripotent cells are very restrictive for several reasons and still few diseases have been captured by these ways (Wakayama et al. 2001; Taranger et al. 2005).

Animal models for human diseases have been using for decades. However they have also some limitations such as showing no or only an approximate resemblance to the human disease, differences in physiology and anatomy between animals and humans, no mirroring for cognitive or behavioral defects of neurological diseases, and different genetic background of animals and humans in terms of resulting phenotype of disease-associated mutations (Colman and Dreesen 2009; Saha and Jaenisch 2009).

To overcome these drawbacks, iPSCs offer disease and patient-specific cells with the knowledge of the clinical history of the donor and they can be made with

S. Yildirim, *Induced Pluripotent Stem Cells*, SpringerBriefs in Stem Cells
DOI: 10.1007/978-1-4614-2206-8_5, © The Author(s) 2012

cells taken from all ages even from elderly patients with chronic disease (Dimos et al. 2008). We can summarize advantages of using in modeling human diseases as:

1. Inexhaustible cell source for experimentation,
2. Limitless self-renewing capacity with pluripotent nature,
3. Development of isogenic cell lines,
4. Personalized treatments (Mattis and Svendsen 2011; Unternaehrer and Daley 2011; Zhu et al. 2011).

Although there are still many challenges regarding to their identity, a couple of reports are available so far to overview a disease phenotype in vitro (Ebert et al. 2009; Lee et al. 2009; Raya et al. 2009; Ye et al. 2009b). Since hESC lines display variable outcomes to differentiate into specific lineages (Osafune et al. 2008), multiple iPSC lines generated from a single patient are extremely favorable by having identical genetic background.

5.1 Disease-Specific iPSCs

Disease-specific iPSCs have been generated from individuals with disorders, such as neurodegenerative diseases, including ALS (Dimos et al. 2008), Parkinson's Disease (Park et al. 2008a; Baek et al. 2009; Soldner et al. 2009; Hargus et al. 2010; Swistowski et al. 2010), SMA (Ebert et al. 2009), familial dysautonomia (Lee et al. 2009) and inherited diseases, including adenosine deaminase deficiency-related severe combined immune deficiency (ADA-SCID), Shwachman-Bodian-Diamond syndrome (SBDS), Gaucher disease (GD) type III, Duchenne (DMD) and Becker muscular dystrophy (BMD), Huntington's disease, juvenile-onset, type 1 diabetes mellitus (JDM), Down syndrome (DS)/trisomy 21, the carrier state of Lesch-Nyhan syndrome (Park et al. 2008b) and Fanconi anaemia (Raya et al. 2009). Unternaehrer and Daley (2011) have reported existing disease-specific iPSC lines available so far from neurological, hematological, cardiovascular and other categories (Table 5.1).

5.2 Choosing Cell Sources

In order to generate patient-specific iPSCs, the first step will be derivation of iPSCs from somatic cells of patient. iPSCs can be made with cells taken from patients of all ages with full medical records, more than 5,000 known genetic diseases, whether simple or complex (Colman and Dreesen 2009). There are available human tissues with no ethical or surgical concern, such as fat, blood, biopsy specimens, skin, plugged hair and extracted tooth (Aasen et al. 2008; Sun

Table 5.1 Disease-specific cell lines from Unternaehrer and Daley (2011)

Disease	Reference	Molecular defect	Donor cell	Age, sex of donor	Method	Differentiated to	Disease phenocopied in	Disease phenocopied in differentiated cells	Drug or functional tests
Neurological									
ALS	(Dimos et al. 2008)	Superoxide dismutase L144F dominant allele	Dermal fibroblasts	82, 89 F	RV4	Motor neurons (HB9 +, ISLET1/2 +, ChAT +)	n.a.	n.d.	No
SMA	(Ebert et al. 2009)	Uncharacterized; decreased *SMN1* expression	Fibroblast	3 M	LVOSLN	Neurons and astrocytes, mature motor neurons	n.a.	Yes: reduced size and number of motor neurons, defective synapse formation, with culture	VPA and tobramycin increased total SMN1 protein and gem formation
Parkinson disease	(Park et al. 2008a; Baek et al. 2009; Soldner et al. 2009; Hargus et al. 2010; Swistowski et al. 2010)	Unknown, *LRRK2*, undefined	Dermal fibroblast	71,53, 53,57, 60,60,85, M,F	RV4, LV3,4, iLV, eLV; some factor free, xRV4	Dopaminergic neurons	n.a.	No	Transplant into brains of rats with Parkinson model; rescue of one of three measures of disease
Huntington disease	(Park et al. 2008a)	72 CAG repeats, huntingtin gene	Fibroblast	20 F	RV4	None	n.a.	n.d.	No
Down syndrome	(Park et al. 2008a; Baek et al. 2009)	Trisomy 21	Fibroblast	1 M	RV4	Teratoma	Yes	Yes; reduced microvessel density	No
Fragile X syndrome	(Urbach et al. 2010)	CGG triplet repeat expansion resulting in silencing of *FMR1*	Dermal fibroblast (2 lines), foetal lung fibroblasts (1 line)	4, 29 y/o, 22 week M	RV4	None	Yes; FMR1 silenced	n.d.	No
Familial dysautonomia	(Lee et al. 2009)	Partial skipping of exon 20 of *IKBKAP*, reduced IKAP protein	Dermal fibroblast	10 F	LV4	Central NS, peripheral NS, haematopoietic, endothelial, endodermal	Higher ratio normal: mutant transcripts in iPS than fibroblast	Yes; decreased expression of genes involved in neurogenesis and neuronal differentiation	decreased migration in wound healing assay; kinetin reduced mutant splice form, and increased per cent of differentiating neurons; no improvement in migration

(continued)

Table 5.1 (continued)

Disease	Reference	Molecular defect	Donor cell	Age, sex of donor	Method	Differentiated to	Disease phenocopied in	Disease phenocopied in differentiated cells	Drug or functional tests
Rett syndrome	(Hotta et al. 2009; Marchetto et al. 2010)	Heterozygous mutation in *MECP2*: C916T, 1155del32, C730T, C473T	Fibroblast	3,5,8 F	LV4 EOS, RV4	None, neural progenitor cells n	n.d., no	n.d., yes; reduced number of spines and density of glutamatergic synapse formation	Yes; IGF1, high dose gentamicin treatment led to more glutamatergic synapses; decreased frequency/intensity of spontaneous currents
Haematological									
ADA-SCID	(Park et al. 2008a)	GGG, AGG, exon 7 and Del(GAAGA) exon 10, *ADA* gene	Fibroblast	3 M	RV4	None	n.a.	n.d.	No
Fanconi anaemia	(Raya et al. 2009)	FA-A, FA-D2 corrected	Dermal fibroblasts	Unknown M	RV4, 2 rounds with 2i	Hematopoietic	No (corrected)	No (corrected)	No
Schwachman–Bodian–Diamond syndrome	(Park et al. 2008a)	Multi-factorial	Bone marrow mesenchymal cells	4 M	RV4	None	n.a.	n.d.	No
Sickle cell anemia	(Mali et al. 2008; Somers et al. 2010)	Undefined? SS	Fibroblast	20 foetal week F	1eLV4; eLV4 + T	None	n.d.	n.d.	No
Beta thalassemia	(Ye et al. 2009a)	Homozygous for codon 41/42 4-bp (CTTT) deletion in beta-globin	Dermal fibroblast	Unknown	RV4	Hematopoietic	n.a.	n.d.	No
Polycythemia vera	(Ye et al. 2009b)	Heterozygous *JAK2* 1849G. T	Fibroblast	Unknown	RV4	Hematopoietic	n.a.	Yes; enhanced erythropoiesis	No
Primary myelofibrosis	(Ye et al. 2009b)	Heterozygous *JAK2* 1849G. T	Fibroblast	Unknown	RV4	None	n.d.	n.d.	No
Metabolic fibroblastno									
Lesch-Nyhan syndrome (carrier)	(Park et al. 2008a; Soldner et al. 2009; Khan et al. 2010)	Heterozygosity of *HPRT1*; A. G mutation in exon 3 of *HPRT1*	Fibroblast	34,11 F	ILV4 + N, iLV3; AAV	None	n.a.	n.d.	No

(continued)

Table 5.1 (continued)

Disease	Reference	Molecular defect	Donor cell	Age, sex of donor	Method	Differentiated to	Disease phenocopied in	Disease phenocopied in differentiated cells	Drug or functional tests
Diabetes type I	(Park et al. 2008a; Maehr et al. 2009)	Multi-factorial; unknown	Fibroblast	42,32,30 F,M	RV3,4	Beta-like cells: somatostatin, glucagon, insulin +, glucoseresponsive	n.a.	n.d.	No
Gaucher disease type III	(Park et al. 2008a)	AAC > AGC, exon 9, Ginsertion, nucleotide 84 of cDNA, *GBA* gene	Fibroblast	20 M	RV4	None	n.a.	n.d.	No
A1ATD	(Rashid et al. 2010; Somers et al. 2010)	A1-antitrypsin deficiency: G342 K	Dermal or liver fibroblasts	65, 55, 47, 57, 61, 64, 0.3 yr F,M	RV4, 1eLV4	Hepatocyte-like cells (fetal)	n.d.	Yes; polymer accumulation	No
GSD1a	(Ghodsizadeh et al. 2010; Rashid et al. 2010)	Hepatic glucose-6-phosphate deficiency	Dermal fibroblasts	25, 7 M	RV4	Hepatocyte-like cells (fetal)	n.d.	Yes; hyper-accumulatio n of glycogen	No
FH	(Rashid et al. 2010)	Familial hypercholesterolemia, autosomal dominant LDLR mutation	Fibroblast	NK	RV4	Hepatocyte-like cells (fetal)	n.d.	Yes; impaired LDL incorporation	No
Crigler–Najjar syndrome	(Ghodsizadeh et al. 2010; Rashid et al. 2010)	13 bp deletion, exon 2 of *UGT1A1*; or L413P	Dermal fibroblasts	2 month old, 19, 21 y/o M,F	RV4	Hepatocyte-like cells (fetal)	n.d.	n.d.	No
Hereditary tyrosinemia type 1	(Ghodsizadeh et al. 2010; Rashid et al. 2010)	553T > G V166G in fumarylaceto-acetate hydrolase	Dermal fibroblasts	2 month old, 6 y/o M,F	RV4	Hepatocyte-like cells (fetal)	n.d.	n.d.	No
Progressive familial hereditary cholestasis	(Ghodsizadeh et al. 2010)	Multi-factorial	Dermal fibroblasts	17 F	RV4	Hepatocyte-like cells (fetal)	n.d.	n.d.	No
Hurler syndrome (MPS IH)	(Tolar et al. 2011)	IDUA deficiency; Y167X, W402X; W402X, W402X	Keratinocyte, MStrC	M 1	RV4	Hematopoietic	Yes; higher GAG accumulation	No difference in CD34 or CD45 + cells or colony formation	No
Cardiovascular									

(continued)

Table 5.1 (continued)

Disease	Reference	Molecular defect	Donor cell	Age, sex of donor	Method	Differentiated to	Disease phenocopied in	Disease phenocopied in differentiated cells	Drug or functional tests
LEOPARD Syndrome	(Carvajal-Vergara et al. 2010)	Heterozygous T468 M in *PTPN11*	Fibroblast	25, 34 F,M	RV4	Cardiomyocytes	n.d.	Yes; cardiomyocyte hypertrophy	Antibody microarray: increase EGFR, MEK1 phosphorylation, no pERK response to bFGF
Long QT syndrome	(Moretti et al. 2010)	Dominant R190Q in *KCNQ1*	Dermal fibroblast	8,42 M	RV4	Cardiomyocytes	n.a.	Yes; longer QT in ventricular and atrial myocytes; impaired cell membrane targeting of KCNQ1 protein	Decreased IKs current density
Other categories									
Duchenne muscular dystrophy	(Park et al. 2008a; Tchieu et al. 2010)	Deletion of exon 45–52 or 46–50, dystrophin gene	Dermal fibroblast	47, 13, 6 F carrier, M affected	RV4, 1LV4	None	n.a.	n.d.	No
Becker muscular dystrophy	(Park et al. 2008a)	Unidentified mutation in dystrophin	Fibroblast	38 M	RV4	None	n.a.	n.d.	No
Dyskeratosis congenita (DC)	(Soldner et al. 2009; Agarwal et al. 2010)	del37L in *DKC1*	Fibroblast	7,30 M	RV4, iLV3	None	n.o.	n.d.	No
Cystic fibrosis	(Somers et al. 2010; Warren et al. 2010)	homozygous delta F508 in *CFTR*	Dermal fibroblast	8,21,29,31,33 F,M	1eLV4	None	n.d.	n.d.	No
Scleroderma	(Somers et al. 2010)	Unknown	Dermal fibroblast	47 F	1eLV4	None	n.d.	n.d.	No
Osteogenesis imperfecta	(Khan et al. 2010)	G > A in exon 34 of *COL1A2*	MSC	NK	AAV OSLN	None	n.a.	n.d.	No

Abbreviations: 2i, inhibition of MEK1, GSK3 with PD0325901 and CT99021; A1ATD or AAT, alpha 1-antitrypsin deficiency; ADA-SCID, adenosine deaminase severe combined immune deficiency; ALS, amyotrophic lateral sclerosis; FMR1, fragile X mental retardation 1; FH, familial hypercholesterolemia; GBA, acid beta-glucosidase; L, Lin28; M, c-Myc; MSC, mesenchymal stem cells; MStrC, mesenchymal stromal cell; MPSIH, mucopolysaccharidosis type I/Hurler syndrome; N, Nanog; n.a., not applicable; n.d., not determined; NK, not known; NS, nervous system; O, Oct4; S, Sox2; SMA, spinal muscular atrophy; VPA, valproic acid. Method abbreviations: AAV, adeno-associated virus; LV3, lentiviral OSK; LV4, lentiviral OSKM; RV3, retroviral OSK; RV4, retroviral OSKM; LVOSLN, lentiviral OSLN; iLV, inducible lentivirus; eLV, Cre-excisable lentivirus; x, xeno-free; 1eLV, single polycistronic vector excisable lentivirus; T, SV40 large T antigen et al. 2009; Ye et al. 2009b; Yan et al. 2010). Thus far, in humans, skin fibroblasts and bone marrow mesenchymal cells (Takahashi et al. 2007; Yu et al. 2007; Huangfu et al. 2008), keratinocytes (Aasen et al. 2008; Maherali et al. 2008), peripheral blood cells (Ye et al. 2009b; Loh et al. 2010), melanocytes (Utikal et al. 2009), neural stem cells (Kim et al. 2009), amniotic fluid-derived cells (Li et al. 2009), adipose stem cells from lipoaspirate (Sun et al. 2009), dental stem cells (Tamaoki et al. 2010; Yan et al. 2010) and mesenchymal stem cells from umbilical cord matrix and amniotic membrane (Cai et al. 2010) have been used for reprogramming. Moreover, frozen-banked tissues, and cell lines can also be used although they supply very little clinical information on the specific donor (Colman and Dreesen 2009).

Since the most disease phenotypes are only observed in differentiated cells, only iPSCs generation could provide a source for pluripotency. Monogenic diseases are the most fruitful targets because the gene and often its product are known. Extending this experimental paradigm to diseases with either unknown or more complex, multifactorial phenotypes or diseases and to model disease with a long latency such as Alzheimer's or Parkinson's Disease would be challenging (Colman and Dreesen 2009). For the diseases that exhibit a late onset in humans, kinetics of disease pathology could be stimulated in the cell culture dish by exposing the cells to in vitro experimental stress (e.g. serum starvation, O_2 reduction, heat shock, etc.) (Saha and Jaenisch 2009). On the other hand, Colman and Dreesen (2009) have emphasized that "late onset" term does not reflect subclinical developments that may occur a lot earlier and may be captured by the in vitro methodology.

Taken together, it is obvious that the disease pattern will direct to alternatives for donor cell type. Since the detection of the mutations from the diseases such as SMA could be possible in all cell types of the patient, skin biopsies can be used as readily accessible donor cells. On the other hand, heterozygous genotype of most of the hematopoietic disorders could be detected only in particular progenitor. Hence, those progenitors should be chosen as cell sources for reprogramming (Ebert et al. 2009; Saha and Jaenisch 2009; Ye et al. 2009b).

Moreover, recent reports about persistent epigenetic imprinting in iPSCs (Kim et al. 2010; Polo et al. 2010) would be useful for an opportunity to researches on sporadic and multifactorial diseases. Although those persistent imprinting manifests as differential gene expression and alters differentiation capacity, it may be utilized in potential therapeutic applications to enhance differentiation into desired cell lineages (Kim et al. 2010; Polo et al. 2010). Especially for the diseases that have a background of combination of genetic and environmental factors, persistent epigenetic memory would be an advantage. In those diseases any epigenetic alterations would be studied via iPSCs carrying parental imprinting. The cells derived from the patients suffering from same disease but living in different geographical regions could give us important clues for environmental factors, such as toxic metals and pesticides, life styles and dietary habits which may effect the epigenome and reflect risk factors (Jaenisch and Bird 2003). Lastly, for the non-cell autonomous diseases, possible success with one cell type may affect the other pathological mediators (Colman and Dreesen 2009).

5.3 Identification of iPSC Colonies

There are several methods or techniques to select or distinguish successfully reprogrammed clones from partially reprogrammed or simply transformed colonies (Stadtfeld and Hochedlinger 2010):

1. The reactivation of endogenous pluripotency-associated genes linked to drug selection cassettes. Fbxo15, Nanog or Oct4 and Klf4 (Takahashi and Yamanaka 2006; Maherali et al. 2007; Okita et al. 2007; Wernig et al. 2007; Pfannkuche et al. 2009),
2. Lentiviral vector systems carrying promoter fragments of pluripotency genes (Hotta et al. 2009),
3. For human iPSCs, TRA-1-81 expression (Lowry et al. 2008),
4. "Indicator retroviruses" expressing fluorescent proteins, which become silenced upon acquisition of pluripotency (Chan et al. 2009),
5. Using only morphological criteria (Blelloch et al. 2007; Maherali et al. 2007; Meissner et al. 2007),
6. The combination of morphological selection and doxycycline-inducible vectors (Stadtfeld et al. 2008),
7. For mouse iPSCs, GH2 gene expression (Liu et al. 2010; Stadtfeld et al. 2010),
8. MEG-3 gene as human homolog of GH2 to identify human iPSCs (Stadtfeld and Hochedlinger 2010).

The generation of patient/disease-specific iPSCs follows standard methods. In brief, target cells are infected with reprogramming system carrying reprogramming factors. After several days, infected cells are trypsinized and replated onto feeder layer cells and media is replaced with standard human ESC media the next day and changed every day thereafter. When appeared, human ESC like colonies are picked mechanically or selected by the systems or methods described above and is passaged. Then pluripotency is evaluated according to the similarity of putative to ESCs (Fig. 5.1). To confirm proper and complete reprogramming, activity of cell cycle profile, normal karyotype maintenance, alkaline phosphatase activity and expression of several ESC-associated antigens (SSEA-3, SSEA-4, TRA1-81, Nanog, etc.), downregulation or lack of immunreactivity against parental cell specific factors (such as fibroblast-associated antigen TE-7), expression of pluripotency genes (Rex1/Zfp42, Foxd3, Tert, Nanog and Cripto/Tdgf1) should be checked. Reverse-transcription PCR and immunocytochemistry can be used to evaluate specific genes or proteins, whereas further examination of the global expression with a microarray gene analysis is useful (Yoshida and Yamanaka 2010).

Generally embryoid bodies (EBs) are created as a first step by aggregating or placing clumps in suspension culture for the evaluation of in vitro differentiation. The resulting EBs are plated onto plastic gelatin coated dishes and allowed to attach for the outgrowth culture. Thereafter iPSC lines will spontaneously differentiate into cell types representative of the three embryonic germ layers (Maherali et al. 2008).

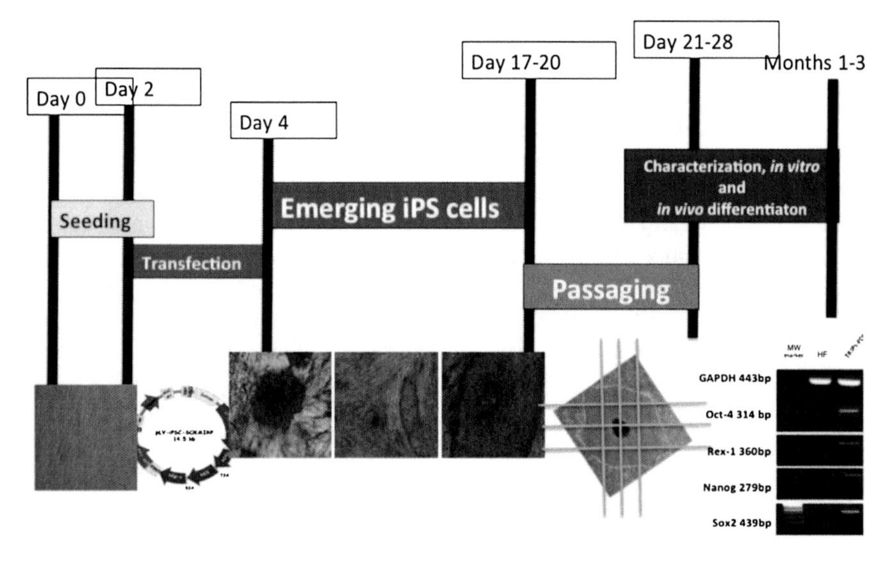

Fig. 5.1 The reprogramming process in dental pulp cells

To determine pluripotency in vivo, iPSCs should be injected into immunocompromised NOD-SCID mice for teratoma assay. Histological analyses of the resulting teratomas should show cell types representative of the three germ layers, including for example pigmented cells for ectodermal, lung, respiratory and gut-like epithelia for endodermal, and mesenchyme, adipose tissue and cartilage for mesodermal differentiation. Immunocytochemistry analyses can be used to detect expression of α-smooth muscle actin (α-SMA), desmin, vimentin, etc. for mesoderm, a-fetoprotein (AFP) for endoderm, glia fibrillary acidic protein (GFAP) and βIII-tubulin for ectoderm markers (Carvajal-Vergara et al. 2010).

The most stringent criterion for mESC/iPSCs is their ability to generate germline competent adult mouse chimeras and thus germline transmission. For human iPSCs, teratoma formation in immune-deficient mice seems to be acceptable so far (Daley et al. 2009; Yoshida and Yamanaka 2010). Besides, the proposed stringent criteria for miPSCs may not be fully required for applications in which reprogrammed human cell lines are used to model disease processes in vitro, and to screen for novel drugs or drug toxicity (Ebert et al. 2009). On the other hand, since current teratoma assays are qualitative in nature, full quantitative assessment of differentiation capacity of generated cell lines could rapidly allow whether those lines retain the capacity for differentiation into three germ layers or had restricted differentiation capacity, even with the lines that fail functional level of forming teratoma. A group of researchers support that the most desirable iPSCs to use for transplantation might be those that do not form teratomas in vivo but retain the capacity to differentiate to desired cell types in vitro. They proposed that this possibility can only be determined if non-teratoma-forming iPSC lines are fully studied in vitro (Ellis et al. 2009). In conclusion, for the purposes of either regenerative medicine or disease modeling, the cells do not have to be germline- or teratoma-competent, as long as they have the ability to self-renewal and differentiation into the desired target cells (Yamanaka 2009a, b).

It was unclear whether reprogramming of female human cells reactivates the inactive X chromosome, as in mouse. It has been recently shown that human iPSCs derived from several female fibroblasts carry inactive X chromosome in contrast to mouse iPSCs that carry two active X chromosomes. While those data indicate that reversal of X chromosome inactivation is not required for human cell programming, the implications that X chromosome inactivation should take into consideration for the use of female iPSCs with devastating X-linked genetic diseases, such as fragile X syndrome (mutation in FMR1), a-thalassemia (ATRX), Rett Syndrome (MECP2), Coffin-Lowry Syndrome (RSK2), DMD, Lesch-Nyhan syndrome (HPRT) and Wiskott-Aldrich Syndrome (WASP) (Tchieu et al. 2010).

5.4 Characterization of Genetic Mutation

The generation of iPSCs from patients with a variety of genetic diseases offers an opportunity to recapitulate both normal and pathological human tissue formation in vitro, hence provide a great tool for disease pathology investigations. If the cells taken from patients having classical Mendelian inherited disorders, point mutations in known genes essential for the given function, molecular mutation analysis such as karyotype analysis and fingerprinting analysis should be carried out (Park et al. 2008b).

To verify that the patient-specific iPSC lines are genetically matched to the donor cells, DNA fingerprinting analysis of the iPSC lines and the donor cells from which they were derived should be done. Additionally, direct sequencing and an allele-specific restriction fragment length polymorphism should be used to compare genotype of iPSC lines with that of the donated host cells. Furthermore, polymerase chain reaction (PCR) analysis of genomic DNA from iPSC lines will reveal that if they carry integrated copies of the transgenes which they have been transduced (Dimos et al. 2008).

5.5 iPSCs Differentiation into Desired Cell Types

Beginning with the first cell lineage decisions, gene expression and mutual interactions between lineage-determining transcription factors with antagonizing functions show stochasticity (Hemberger et al. 2009). The orchestrating role of the gene regulatory network points biological patterns for differentiation. To date, ESC studies have mostly focused on the derivation of subsets of tissue-specific cell populations. Thus, lineage specific differentiation of murine and human ESCs has been shown as a powerful tool for studying early embryonic events and lineage restriction for generating an unlimited cellular supply for cell therapies and tissue engineering. To produce either progenitors or more mature cells, various exogenous factors were applied in a sequence and time course, which is highly reminiscent of normal development.

Much of the hope invested in patient-specific stem cells is based on the assumption that it will be possible to differentiate them into disease relevant cell types. To differentiate desired cell types within the mixed population of differentiating ESCs, there are several well-established stem cell differentiation protocols mimicking the proper timeline of normal human organogenesis. For hematopoietic, endothelial, osteoblastic, osteoclastic (Tsuneto et al. 2003; Kawaguchi et al. 2005; Grigoriadis et al. 2010), cardiac (Doetschman et al. 1985; Schenke-Layland et al. 2008; Arbel et al. 2010), neural commitment (Ying and Smith 2003), pancreatic (Banerjee et al. 2011), hepatic (Gerbal-Chaloin et al. 2010), chondrogenic and adipogenic differentiations (Wdziekonski et al. 2003) many very well-established protocols have been used. Moreover, cell fate specification and maturation could be selectively altered via manipulation of endogenous developmental signaling pathways (Rathjen and Rathjen 2003; Meyer et al. 2009). The ability to differentiate iPSCs in vitro to specific lineages efficiently and reproducibly has been achieved by using the described protocols with certain modifications (Dimos et al. 2008; Narazaki et al. 2008; Tateishi et al. 2008; Wernig et al. 2008a; Ebert et al. 2009; Hu and Zhang 2009; Jin et al. 2009; Karumbayaram et al. 2009; Pfannkuche et al. 2009; Senju et al. 2009; Tanaka et al. 2009; Taura et al. 2009; Zhang et al. 2009; Carvajal-Vergara et al. 2010; Comyn et al. 2010; Dick et al. 2010; Gamm and Meyer 2010; Huang et al. 2010; Kaichi et al. 2010; Lamba et al. 2010; Lee et al. 2010; Martinez-Fernandez et al. 2010; Parameswaran et al. 2010; Swistowski et al. 2010; Teramura et al. 2010; Zhou et al. 2010).

There are many typical examples for directed differentiation of iPSCs to the cell types influenced by the disease in recent publications (Tateishi et al. 2008; Meyer et al. 2009; Grigoriadis et al. 2010). In those studies the researchers showed that it is possible to generate and fine-tune desired lineages from human somatic cells. In brief, after creating EBs from generated iPSCs, chemically defined differentiation media were used to promote the stepwise production of organ-specific cell types. Therefore using targeted, stepwise differentiation process that follows a normal developmental timeline, the researcher modeled cell and/or organ development with human iPSCs (Ueda et al. 2010).

Differentiation should be confirmed by showing the expressions of transcription factors or surface markers. Apparently, the function of those differentiated cells will be the last point for evaluation (Dimos et al. 2008). Dimos and colleagues demonstrated that skin cells from patients with amyotrophic lateral sclerosis, ALS, could be reprogrammed and subsequently differentiated into disease-free motor neurons (Dimos et al. 2008). Also of note, Ebert et al. (2009) made iPSCs from the fibroblasts of a SMA patient and his unaffected mother. They were the first to demonstrate a preserved patient-specific disease phenotype in motor neurons generated from fibroblast iPSCs. Treatment of these cells in vitro with valproic acid and tobramycin led to an upregulation of survival motor neuron protein synthesis and displayed selective deficits when compared with normal motor neurons.

Today, many iPSC lines could be directed into a differentiated functional cell types. Auditory retinal cells (Jin et al. 2009; Comyn et al. 2010; Lamba et al. 2010; Parameswaran et al. 2010), cardiomyocytes (Narazaki et al. 2008; Pfannkuche et al. 2009; Tanaka et al. 2009; Zhang et al. 2009; Carvajal-Vergara et al. 2010; Kaichi et al. 2010; Martinez-Fernandez et al. 2010), insulin-secreting islet-like clusters (Tateishi et al. 2008), motor neurons (Dimos et al. 2008; Ebert et al. 2009; Hu and Zhang 2009; Karumbayaram et al. 2009), dopaminergic neurons (Wernig et al. 2008a; Cai et al. 2010; Swistowski et al. 2010), auditory spinal ganglion neurons (Nishimura et al. 2009), smooth muscle cells (Taura et al. 2009; Xie et al. 2009), vascular endothelial cells (Taura et al. 2009), dendritic cells and macrophages (Senju et al. 2009; Senju et al. 2010), adipocytes (Tashiro et al. 2009; Taura et al. 2009), osteoblasts (Tashiro et al. 2009), hematopoietic cells (Kauffman 1993; Lu et al. 2009; Okabe et al. 2009; Raya et al. 2009; Kaneko et al. 2010), endothelial progenitor cells (Xu et al. 2009; Abaci et al. 2010; Alipio et al. 2010; Ho et al. 2010; Homma et al. 2010) are available.

Furthermore, Hanna et al. have shown that differentiated iPSCs can be used to rescue organ function in humanized mouse model of sickle cell anemia (Hanna et al. 2007). More recently several other groups have demonstrated the therapeutic potential of iPSCs, both alone and in combination with genetic corrective therapy (Table 5.1). They include the generation of disease-free hematopoietic progenitors from keratinocytes obtained from patients with Fanconi anemia (Raya et al. 2009), the correction of hemophilia in mice using iPSC-derived endothelial progenitors (Xu et al. 2009), and multilineage functional repair of the diseased heart tissue in immunocompetent mice using undifferentiated iPSCs (Nelson et al. 2009). Finally, functional dopamine neurons could be generated from reprogrammed mouse fibroblasts, and transplantation of these neurons could restore dopamine function when grafted in Parkinsonian rats (Wernig et al. 2008a).

References

Aasen T et al (2008) Efficient and rapid generation of induced pluripotent stem cells from human keratinocytes. Nat Biotechnol 26(11):1276–1284

Abaci HE et al (2010) Adaptation to oxygen deprivation in cultures of human pluripotent stem cells, endothelial progenitor cells, and umbilical vein endothelial cells. Am J Physiol Cell Physiol 298(6):C1527–C1537

Agarwal S et al (2010) Telomere elongation in induced pluripotent stem cells from dyskeratosis congenita patients. Nature 464(7286):292–296

Alipio Z et al (2010) Sustained factor VIII production in hemophiliac mice 1 year after engraftment with induced pluripotent stem cell-derived factor VIII producing endothelial cells. Blood Coagul Fibrinolysis 21(5):502–504

Arbel G et al (2010) Methods for human embryonic stem cells derived cardiomyocytes cultivation, genetic manipulation, and transplantation. Methods Mol Biol 660:85–95

Baek KH et al (2009) Down's syndrome suppression of tumour growth and the role of the calcineurin inhibitor DSCR1. Nature 459(7250):1126–1130

Banerjee I et al (2011) Impact of co-culture on pancreatic differentiation of embryonic stem cells. J Tissue Eng Regen Med 5(4):313–323

Blelloch R et al (2007) Generation of induced pluripotent stem cells in the absence of drug selection. Cell Stem Cell 1(3):245–247
Cai J et al (2010) Dopaminergic neurons derived from human induced pluripotent stem cells survive and integrate into 6-OHDA-lesioned rats. Stem Cells Dev 19(7):1017–1023
Carvajal-Vergara X et al (2010) Patient-specific induced pluripotent stem-cell-derived models of LEOPARD syndrome. Nature 465(7299):808–812
Chan EM et al (2009) Live cell imaging distinguishes bona fide human iPS cells from partially reprogrammed cells. Nat Biotechnol 27(11):1033–1037
Colman A, Dreesen O (2009) Pluripotent stem cells and disease modeling. Cell Stem Cell 5(3):244–247
Comyn O et al (2010) Induced pluripotent stem cell therapies for retinal disease. Curr Opin Neurol 23(1):4–9
Daley GQ et al (2009) Broader implications of defining standards for the pluripotency of iPSCs. Cell Stem Cell 4(3):200–201; author reply 202
Dick E et al (2010) Evaluating the utility of cardiomyocytes from human pluripotent stem cells for drug screening. Biochem Soc Trans 38(4):1037–1045
Dimos JT et al (2008) Induced pluripotent stem cells generated from patients with ALS can be differentiated into motor neurons. Science 321(5893):1218–1221
Doetschman TC et al (1985) The in vitro development of blastocyst-derived embryonic stem cell lines: formation of visceral yolk sac, blood islands and myocardium. J Embryol Exp Morphol 87:27–45
Ebert AD et al (2009) Induced pluripotent stem cells from a spinal muscular atrophy patient. Nature 457(7227):277–280
Ellis J et al (2009) Alternative induced pluripotent stem cell characterization criteria for in vitro applications. Cell Stem Cell 4(3):198–199; author reply 202
Gamm DM, Meyer JS (2010) Directed differentiation of human induced pluripotent stem cells: a retina perspective. Regen Med 5(3):315–317
Gearhart J (1998) New potential for human embryonic stem cells. Science 282(5391):1061–1062
Gerbal-Chaloin S et al (2010) Isolation and culture of adult human liver progenitor cells: in vitro differentiation to hepatocyte-like cells. Methods Mol Biol 640:247–260
Ghodsizadeh A et al (2010) Generation of liver disease-specific induced pluripotent stem cells along with efficient differentiation to functional hepatocyte-like cells. Stem Cell Rev 6(4):622–632
Grigoriadis AE et al (2010) Directed differentiation of hematopoietic precursors and functional osteoclasts from human ES and iPS cells. Blood 115(14):2769–2776
Hanna J et al (2007) Treatment of sickle cell anemia mouse model with iPS cells generated from autologous skin. Science 318(5858):1920–1923
Hargus G et al (2010) Differentiated Parkinson patient-derived induced pluripotent stem cells grow in the adult rodent brain and reduce motor asymmetry in Parkinsonian rats. Proc Natl Acad Sci USA 107(36):15921–15926
Hemberger M et al (2009) Epigenetic dynamics of stem cells and cell lineage commitment: digging Waddington's canal. Nat Rev Mol Cell Biol 10(8):526–537
Ho PJ et al (2010) Endogenous KLF4 expression in human fetal endothelial cells allows for reprogramming to pluripotency with just OCT3/4 and SOX2—brief report. Arterioscler Thromb Vasc Biol 30(10):1905–1907
Homma K et al (2010) Sirtl plays an important role in mediating greater functionality of human ES/iPS-derived vascular endothelial cells. Atherosclerosis 212(1):42–47
Hotta A et al (2009) Isolation of human iPS cells using EOS lentiviral vectors to select for pluripotency. Nat Methods 6(5):370–376
Hu BY, Zhang SC (2009) Differentiation of spinal motor neurons from pluripotent human stem cells. Nat Protoc 4(9):1295–1304
Huang HP et al (2010) Factors from human embryonic stem cell-derived fibroblast-like cells promote topology-dependent hepatic differentiation in primate embryonic and induced pluripotent stem cells. J Biol Chem 285(43):33510–33519

Huangfu D et al (2008) Induction of pluripotent stem cells from primary human fibroblasts with only Oct4 and Sox2. Nat Biotechnol 26(11):1269–1275

Jaenisch R, Bird A (2003) Epigenetic regulation of gene expression: how the genome integrates intrinsic and environmental signals. Nat Genet 33(suppl):245–254

Jin ZB et al (2009) Induced pluripotent stem cells for retinal degenerative diseases: a new perspective on the challenges. J Genet 88(4):417–424

Kaichi S et al (2010) Cell line-dependent differentiation of induced pluripotent stem cells into cardiomyocytes in mice. Cardiovasc Res 88(2):314–323

Kaneko S et al (2010) Reprogramming adult hematopoietic cells. Curr Opin Hematol 17(4):271–275

Karumbayaram S et al (2009) Directed differentiation of human-induced pluripotent stem cells generates active motor neurons. Stem Cells 27(4):806–811

Kauffman SA (1993) Self-Organization and Adaptation in Complex System. Oxford University Press, Oxford

Kawaguchi J et al (2005) Osteogenic and chondrogenic differentiation of embryonic stem cells in response to specific growth factors. Bone 36(5):758–769

Khan IF et al (2010) Engineering of human pluripotent stem cells by AAV-mediated gene targeting. Mol Ther 18(6):1192–1199

Kim JB et al (2009) Direct reprogramming of human neural stem cells by OCT4. Nature 461(7264):649–653

Kim K et al (2010) Epigenetic memory in induced pluripotent stem cells. Nature 467(7313):285–290

Lamba DA et al (2010). Generation, purification and transplantation of photoreceptors derived from human induced pluripotent stem cells. PLoS One 5(1):e8763

Lee G et al (2009) Modelling pathogenesis and treatment of familial dysautonomia using patient-specific iPSCs. Nature 461(7262):402–406

Lee G et al (2010) Derivation of neural crest cells from human pluripotent stem cells. Nat Protoc 5(4):688–701

Li W et al (2009) Generation of rat and human induced pluripotent stem cells by combining genetic reprogramming and chemical inhibitors. Cell Stem Cell 4(1):16–19

Liu H et al (2010) Generation of endoderm-derived human induced pluripotent stem cells from primary hepatocytes. Hepatology 51(5):1810–1819

Loh YH et al (2010) Reprogramming of T cells from human peripheral blood. Cell Stem Cell 7(1):15–19

Lowry WE et al (2008) Generation of human induced pluripotent stem cells from dermal fibroblasts. Proc Natl Acad Sci USA 105(8):2883–2888

Lu M et al (2009) Enhanced generation of hematopoietic cells from human hepatocarcinoma cell-stimulated human embryonic and induced pluripotent stem cells. Exp Hematol 37(8):924–936

Maehr R et al (2009) Generation of pluripotent stem cells from patients with type 1 diabetes. Proc Natl Acad Sci USA 106(37):15768–15773

Maherali N et al (2007) Directly reprogrammed fibroblasts show global epigenetic remodeling and widespread tissue contribution. Cell Stem Cell 1(1):55–70

Maherali N et al (2008) A high-efficiency system for the generation and study of human induced pluripotent stem cells. Cell Stem Cell 3(3):340–345

Mali P et al (2008) Improved efficiency and pace of generating induced pluripotent stem cells from human adult and fetal fibroblasts. Stem Cells 26(8):1998–2005

Marchetto MC et al (2010) A model for neural development and treatment of Rett syndrome using human induced pluripotent stem cells. Cell 143(4):527–539

Martinez-Fernandez A et al (2010) c-MYC independent nuclear reprogramming favors cardiogenic potential of induced pluripotent stem cells. J Cardiovasc Transl Res 3(1):13–23

Mattis VB, Svendsen CN (2011) Induced pluripotent stem cells: a new revolution for clinical neurology? Lancet Neurol 10(4):383–394

Meissner A et al (2007) Direct reprogramming of genetically unmodified fibroblasts into pluripotent stem cells. Nat Biotechnol 25(10):1177–1181

Moretti A et al (2010) Patient-specific induced pluripotent stem-cell models for long-QT syndrome. N Engl J Med 363(15):1397–1409

Meyer JS et al (2009) Modeling early retinal development with human embryonic and induced pluripotent stem cells. Proc Natl Acad Sci USA 106(39):16698–16703
Narazaki G et al (2008) Directed and systematic differentiation of cardiovascular cells from mouse induced pluripotent stem cells. Circulation 118(5):498–506
Nelson TJ et al (2009) Repair of acute myocardial infarction by human stemness factors induced pluripotent stem cells. Circulation 120(5):408–416
Nishimura K et al (2009) Transplantation of mouse induced pluripotent stem cells into the cochlea. Neuroreport 20(14):1250–1254
Okabe M et al (2009) Definitive proof for direct reprogramming of hematopoietic cells to pluripotency. Blood 114(9):1764–1767
Okita K et al (2007) Generation of germline-competent induced pluripotent stem cells. Nature 448(7151):313–317
Osafune K et al (2008) Marked differences in differentiation propensity among human embryonic stem cell lines. Nat Biotechnol 26(3):313–315
Parameswaran S et al (2010) Induced pluripotent stem cells generate both retinal ganglion cells and photoreceptors: therapeutic implications in degenerative changes in glaucoma and age-related macular degeneration. Stem Cells 28(4):695–703
Park IH et al (2008a) Disease-specific induced pluripotent stem cells. Cell 134(5):877–886
Park IH et al (2008b) Disease-specific induced pluripotent stem cells. Cell 134(5):877–886
Pfannkuche K et al (2009) Cardiac myocytes derived from murine reprogrammed fibroblasts: intact hormonal regulation, cardiac ion channel expression and development of contractility. Cell Physiol Biochem 24(1–2):73–86
Polo JM et al (2010) Cell type of origin influences the molecular and functional properties of mouse induced pluripotent stem cells. Nat Biotechnol 28(8):848–855
Rathjen J, Rathjen PD (2003) Lineage specific differentiation of mouse ES cells: formation and differentiation of early primitive ectoderm-like (EPL) cells. Methods Enzymol 365:3–25
Rashid ST et al (2010) Modeling inherited metabolic disorders of the liver using human induced pluripotent stem cells. J Clin Invest 120(9):3127–3136
Raya A et al (2009) Disease-corrected haematopoietic progenitors from Fanconi anaemia induced pluripotent stem cells. Nature 460(7251):53–59
Saha K, Jaenisch R (2009) Technical challenges in using human induced pluripotent stem cells to model disease. Cell Stem Cell 5(6):584–595
Schenke-Layland K et al (2008) Reprogrammed mouse fibroblasts differentiate into cells of the cardiovascular and hematopoietic lineages. Stem Cells 26(6):1537–1546
Senju S et al (2009) Characterization of dendritic cells and macrophages generated by directed differentiation from mouse induced pluripotent stem cells. Stem Cells 27(5):1021–1031
Senju S et al (2010) Pluripotent stem cells as source of dendritic cells for immune therapy. Int J Hematol 91(3):392–400
Soldner F et al (2009) Parkinson's disease patient-derived induced pluripotent stem cells free of viral reprogramming factors. Cell 136(5):964–977
Somers A et al (2010) Generation of transgene-free lung disease-specific human induced pluripotent stem cells using a single excisable lentiviral stem cell cassette. Stem Cells 28(10):1728–1740
Stadtfeld M, Hochedlinger K (2010) Induced pluripotency: history, mechanisms, and applications. Genes Dev 24(20):2239–2263
Stadtfeld M et al (2008) Defining molecular cornerstones during fibroblast to iPS cell reprogramming in mouse. Cell Stem Cell 2(3):230–240
Stadtfeld M et al (2010) A reprogrammable mouse strain from gene-targeted embryonic stem cells. Nat Methods 7(1):53–55
Sun N et al (2009) Feeder-free derivation of induced pluripotent stem cells from adult human adipose stem cells. Proc Natl Acad Sci USA 106(37):15720–15725
Swistowski A et al (2010) Efficient Generation of Functional Dopaminergic Neurons from Human Induced pluripotent Stem Cells under Defined Conditions. Stem Cells 28(10):1893–1904

Takahashi K, Yamanaka S (2006) Induction of pluripotent stem cells from mouse embryonic and adult fibroblast cultures by defined factors. Cell 126(4):663–676
Takahashi K et al (2007) Induction of pluripotent stem cells from adult human fibroblasts by defined factors. Cell 131(5):861–872
Tamaoki N et al (2010) Dental pulp cells for induced pluripotent stem cell banking. J Dent Res 89(8):773–778
Tanaka T et al (2009) In vitro pharmacologic testing using human induced pluripotent stem cell-derived cardiomyocytes. Biochem Biophys Res Commun 385(4):497–502
Taranger CK et al (2005) Induction of dedifferentiation, genomewide transcriptional programming, and epigenetic reprogramming by extracts of carcinoma and embryonic stem cells. Mol Biol Cell 16(12):5719–5735
Tashiro K et al (2009) Efficient adipocyte and osteoblast differentiation from mouse induced pluripotent stem cells by adenoviral transduction. Stem Cells 27(8):1802–1811
Tateishi K et al (2008) Generation of insulin-secreting islet-like clusters from human skin fibroblasts. J Biol Chem 283(46):31601–31607
Taura D et al (2009) Adipogenic differentiation of human induced pluripotent stem cells: comparison with that of human embryonic stem cells. FEBS Lett 583(6):1029–1033
Tchieu J et al (2010) Female Human iPSCs Retain an Inactive X Chromosome. Cell Stem Cell 7(3):329–342
Teramura T et al (2010) Induction of mesenchymal progenitor cells with chondrogenic property from mouse-induced pluripotent stem cells. Cell Reprogram 12(3):249–261
Thomson JA et al (1998) Embryonic stem cell lines derived from human blastocysts. Science 282(5391):1145–1147
Tolar J et al (2011) Hematopoietic differentiation of induced pluripotent stem cells from patients with mucopolysaccharidosis type I (Hurler syndrome). Blood 117(3):839–847
Tsuneto M et al (2003) In vitro differentiation of mouse ES cells into hematopoietic, endothelial, and osteoblastic cell lineages: the possibility of in vitro organogenesis. Methods Enzymol 365:98–114
Ueda T et al (2010) Generation of functional gut-like organ from mouse induced pluripotent stem cells. Biochem Biophys Res Commun 391(1):38–42
Unternaehrer JJ, Daley GQ (2011) Induced pluripotent stem cells for modelling human diseases. Philos Trans R Soc Lond B Biol Sci 366(1575):2274–2285
Urbach A et al (2010) Differential modeling of fragile X syndrome by human embryonic stem cells and induced pluripotent stem cells. Cell Stem Cell 6(5):407–411
Utikal J et al (2009) Sox2 is dispensable for the reprogramming of melanocytes and melanoma cells into induced pluripotent stem cells. J Cell Sci 122(Pt 19):3502–3510
Wakayama T et al (2001) Differentiation of embryonic stem cell lines generated from adult somatic cells by nuclear transfer. Science 292(5517):740–743
Warren L et al (2010) Highly efficient reprogramming to pluripotency and directed differentiation of human cells with synthetic modified mRNA. Cell Stem Cell 7(5):618–630
Wdziekonski B et al (2003) Development of adipocytes from differentiated ES cells. Methods Enzymol 365:268–277
Wernig M et al (2007) In vitro reprogramming of fibroblasts into a pluripotent ES-cell-like state. Nature 448(7151):318–324
Wernig M et al (2008) A drug-inducible transgenic system for direct reprogramming of multiple somatic cell types. Nat Biotechnol 26(8):916–924
Xie CQ et al (2009) A comparison of murine smooth muscle cells generated from embryonic versus induced pluripotent stem cells. Stem Cells Dev 18(5):741–748
Xu D et al (2009) Phenotypic correction of murine hemophilia A using an iPS cell-based therapy. Proc Natl Acad Sci USA 106(3):808–813
Yamanaka S (2009a) Elite and stochastic models for induced pluripotent stem cell generation. Nature 460(7251):49–52
Yamanaka S (2009b) A fresh look at iPS cells. Cell 137(1):13–17

Yan X et al (2010) iPS cells reprogrammed from human mesenchymal-like stem/progenitor cells of dental tissue origin. Stem Cells Dev 19(4):469–480

Ye L et al (2009a) Induced pluripotent stem cells offer new approach to therapy in thalassemia and sickle cell anemia and option in prenatal diagnosis in genetic diseases. Proc Natl Acad Sci USA 106(24):9826–9830

Ye Z et al (2009b) Human-induced pluripotent stem cells from blood cells of healthy donors and patients with acquired blood disorders. Blood 114(27):5473–5480

Ying QL, Smith AG (2003) Defined conditions for neural commitment and differentiation. Methods Enzymol 365:327–341

Yoshida Y, Yamanaka S (2010) Recent stem cell advances: induced pluripotent stem cells for disease modeling and stem cell-based regeneration. Circulation 122(1):80–87

Yu J et al (2007) Induced pluripotent stem cell lines derived from human somatic cells. Science 318(5858):1917–1920

Zhang J et al (2009) Functional cardiomyocytes derived from human induced pluripotent stem cells. Circ Res 104(4):e30–e41

Zhou J et al (2010) High-Efficiency Induction of Neural Conversion in hESCs and hiPSCs with a Single Chemical Inhibitor of TGF-beta Superfamily Receptors. Stem Cells 28(10):1741–1750

Zhu H et al (2011) Investigating monogenic and complex diseases with pluripotent stem cells. Nat Rev Genet 12(4):266–275

Chapter 6
Challenges to Therapeutic Potential of hiPSCs

In a metazoan body all cells possess the same set of genes. Exceptions for this condition are post-meiotic germ cell lines, mature lymphocytes and cells in species that exhibit chromosome diminution (Kloc and Zagrodzinska 2001). Therefore, generating a pluripotent cell in vitro and directing its conversion into a specific differentiated cell fate, which means rewinding the internal clock of any mammalian cell to an embryonic state and then forwarding this high potential cell to diseased cells, represents a rational and ongoing approach in regenerative medicine. On the other hand, quality control and safety are the main concerns and there are several technical challenges in using human iPSCs in treatment of several irreparable human diseases. To minimize or eliminate genetic alterations in the derived iPSC line creation factor-free human iPSCs are necessary. Defining a disease-relevant phenotype needs in vitro and in vivo models. Moreover, to generate markers for differentiation and gene corrections, gene-targeting strategies are necessary. Besides, cell-type specific lineage reporters, lineage-tracking tools and tools to disrupt, repair or overexpress genes should be developed in order to model many human diseases (Saha and Jaenisch 2009).

Since cellular functions are influenced by microenvironmental stimuli, it is important to evaluate the results obtained from iPSC studies regarding the reprogramming method, culture conditions and differentiation protocols, which all influence the outcome (Daley et al. 2009). Kim et al. (2010) revealed several important facts about reprogramming and resultant iPSCs: first, tissue source influences the efficiency and fidelity of reprogramming. Second, there are substantial differences between iPS and embryo-derived ES cells. Third, the differentiation propensity and methylation profile of iPSCs could be reset. Finally and most strikingly, NT-derived ESCs are more faithfully reprogrammed than most iPSCs generated from adult somatic tissues (Kim et al. 2010).

In addition, in order to determine whether derived iPSCs are specific to a given cell line or to a pluripotency, healthy wild-type controls should be used. Although established human ESC or iPSC lines could be used for this purpose, additionally a panel of lines derived from the same patient or unrelated patients suffering from

S. Yildirim, *Induced Pluripotent Stem Cells*, SpringerBriefs in Stem Cells
DOI: 10.1007/978-1-4614-2206-8_6, © The Author(s) 2012

the same disease would give valuable information. On the other hand, in single gene diseases, genetically modified iPSCs could represent an ideal isogenic control (Colman and Dreesen 2009; Saha and Jaenisch 2009). However, isogenic generation of iPSC lines would be applied only to diseases with known genetic causes (Mattis and Svendsen 2011).

Although there are still many challenges regarding the identity of iPSCs, a couple of reports are available so far to overview a disease phenotype in vitro (Ebert et al. 2009; Lee et al. 2009; Raya et al. 2009; Ye et al. 2009). Since human ESC lines display variable outcomes in differentiating into specific lineages (Osafune et al. 2008), multiple iPSC lines generated from a single patient are extremely favorable because of providing identical genetic backgrounds. The challenges for reprogramming technology can be summarized as follows:

1. Traditional reprogramming techniques may not be satisfactory,
2. Presence of viral vectors,
3. Availability of reliable and repeatable protocols for complete differentiation to a tissue type of choice,
4. Variability in the ability of differentiating to a population of choice and differentiated cells are generally short lived,
5. Necessity of ESC comparisons/lack of proper control,
6. Aberrations in prolonged cultures,
7. Non-cell-autonomous phenotypes and diseases of aging,
8. Low penetrance and modest or undetectable phenotypes for polygenic or complex factors (Mattis and Svendsen 2011; Zhu et al. 2011).

Lack of immunogenicity has been the biggest hope for using custom-made, patient-specific adult cells derived from iPSCs to treat patients with a wide range of diseases. However, Zhao and colleagues showed very recently that virally or episomally produced iPSCs are immunogenic (Zhao et al. 2011). In this simple but smart experimental approach, the researchers injected autologous ESCs, unmatched ESCs and autologous iPSCs derived from fetal fibroblasts into the matched mice. Surprisingly, iPSCs, like unmatched ESCs were rejected by the immune system. Since the viral vectors used for reprogramming could have been responsible for immune response, using episomal vectors had been used. Even with this setting immune rejection was persistent albeit with a weakened response. Overall the data were unexpected: in teratoma formation assay autologous iPSCs were immunogenic than matched ESCs (Zhao et al. 2011). Furthermore, the authors suggested that T-cell dependent immune response in syngeneic recipients by abnormal gene expression in some cells differentiated from iPSCs might be the reason for the immunogenicity of iPSCs in this study. Gene expression profiles of iPSC-derived teratomas were analyzed and it was found that a group of nine genes was expressed at abnormally high levels. The induced expression of three of those 9 genes (Hormad1, Zg16 and Cyp3a11) prevented teratoma formation by non-immunogenic ESCs. As a result, Zhao et al. (2011) suggested that the expression of these minor antigens could be another proof regarding the epigenetic difference between iPSCs and ESCs.

Although immature iPSCs would never be used for transplantation in clinical settings (Apostolou and Hochedlinger 2011), differentiated cells derived from iPSCs could also be immunogenic. It has been shown that upon transplantation, ESC-derived cells could express additional molecules while they are maturing in vivo. Therefore, they become more immunogenic (Swijnenburg et al. 2005).

It is clear that there are still important issues resisting the accumulating hopeful data regarding the therapeutic usage of stem cells. Taylor et al. (2011) scrutinized the controversy about immunogenicity of stem cell-derived tissues. The authors stated that the documentation of HLA (or MHC) expression by ESCs has shaded the reality that any transplanted stem cell-derived tissues, if they are not genetically identical to the recipient, has the potential to induce allograft rejection via "indirect" recognition (Taylor et al. 2011). Besides, absence of the expression of HLA molecules cannot prevent immunological rejection because of activation of NK cells (Bryceson and Long 2008).

Although it is still unclear whether any of the iPSC lines can be used for future cell therapy, it should be quite useful to establish the clinical-grade iPSC banks with a sufficient repertoire of HLA types. Recently, Nakatsuji (2010) estimated that a collection of unique iPSC lines having homozygous alleles of the 3 HLA loci (A, B, and DR) would enable full matching for 80–90% of the Japanese population with a perfect match (Nakatsuji 2010). In addition, Tamaoki et al. (2010) attempted to use dental pulp stem cells to generate iPSC banking having a sufficient repertoire of HLA types. They also reported the possibility of identifying homozygous donors for human iPSC lines for the construction of such HLA-type banking. The practical isolation and handling of dental pulp cells may make it easy to expand the size of the bank in multiple institutes and even establish a number of iPSC lines homozygous for the 3 HLA loci (Tamaoki et al. 2010). In conclusion, establishing stem cell banks that contain pluripotent cells that are closely matched or compatible for HLA along with the induction of antigen-specific immunological tolerance instead of life-long immunosuppressive therapy are hopes for the avoidance of immunological rejection (Taylor et al. 2011).

6.1 Is Reprogramming Necessary for Regenerative Therapies?

Along with epigenetic discussion another important question is whether it is necessary to reprogram cells back to pluripotent stem cell state or not. For regenerative therapies pluripotency may not be a prerequisite for the generation of certain differentiated cell types.

The experimental transdifferentiation of B cells into macrophages, and pancreatic exocrine cells into insulin producing β-cells provides good examples for direct conversion of one cell type into another (Xie et al. 2004; Zhou et al. 2008).

It has been shown that overexpression or deletion of individual transcription factors could change the cell fate in somatic cells (Hochedlinger and Plath 2009). Moreover, culturing stem cells in defined culture conditions can initiate differentiation programs (Kocaefe et al. 2010). Besides, it has been recently shown that the direct conversion of fibroblasts to functional neurons with no prior pluripotent stage is quite possible (Vierbuchen et al. 2010). Fibroblasts have also been transdifferentiated into cardiomyocytes using a pool of 13 transcription and epigenetic remodeling factors (Ieda et al. 2010). Subsequently, only three (GATA4, MEF2C and TBX5) were found to be enough to drive the genetic program regulating cardiomyocyte differentiation (Lin et al. 1997; Garg et al. 2003; Ghosh et al. 2009).

Masip et al. (2010) described, "induced transdifferentiated (iT) cells," as a novel tool for adult cell fate modification. Interconversion between adult cells from ontogenically different lineages by an induced transdifferentiation process is based on the overexpression of a single or cocktail of transcription factors. Since there is no attempt to reach through an embryonic stem cell-like state, iT cells may provide an alternative for regenerative medicine. On the other hand, like iPS cells, they also need safe methods with transient and/or non-integrative tools for generation. Moreover they also need in vivo assays to determine the suitability of their transplantation and applicability in regenerative medicine (Masip et al. 2010).

In conclusion, as Meissner stated, the identification of putative enhancers, miRNAs, and large intergenic non-coding (linc) RNAs may help us to gain further insights into the transcriptional regulation and the role of epigenetic modifications (Meissner 2010). However, the flowing information coming from the rapidly advancing field of non-coding RNAs clearly shows that we are still far from understanding how the epigenetic machinery operates cells (Meissner 2010). iPSCs opened a new era in regenerative medicine since the most desirable achievement in cell biology research would be getting a genuine regenerative response from a pluripotent cell that can differentiate into almost any cell type. Moreover, iPSCs could break the walls of fundamental hypothesis in cell biology. These cells not only changed the broadly accepted and deeply rooted thoughts about terminally differentiated cells in developmental biology, but also helped us to understand how a cell uses its very existing potential to produce other cell types. By compelling the limits of conventional biology, iPSCs opened a new era for biologists who need more than words some conceivable formulas generating from fundamental laws of mathematics and physics to understand complex and emergent behavior of lineage commitment and pluripotency. There is an emergent field in cell biology that aims to clarify the unfathomably complex interactions of genes by using general concepts or principles of physics and mathematics to establish a firm theoretical foundation. Those quantitative models of cell state transitions give us an opportunity to guide future experimentation aimed at dissecting, and therefore directing the mechanisms of cell transitions (Hanna et al. 2010). We would say, iPSCs changed the minds so radically that we may be ready to move using our theoretical backgrounds in a different way.

References

Apostolou E, Hochedlinger K (2011) Stem cells: iPS cells under attack. Nature 474(7350):165–166

Bryceson YT, Long EO (2008) Line of attack: NK cell specificity and integration of signals. Curr Opin Immunol 20(3):344–352

Colman A, Dreesen O (2009) Pluripotent stem cells and disease modeling. Cell Stem Cell 5(3):244–247

Daley GQ et al (2009) Broader implications of defining standards for the pluripotency of iPSCs. Cell Stem Cell 4(3):200–201; author reply 202

Ebert AD et al (2009) Induced pluripotent stem cells from a spinal muscular atrophy patient. Nature 457(7227):277–280

Garg V et al (2003) GATA4 mutations cause human congenital heart defects and reveal an interaction with TBX5. Nature 424(6947):443–447

Ghosh TK et al (2009) Physical interaction between TBX5 and MEF2C is required for early heart development. Mol Cell Biol 29(8):2205–2218

Hanna JH et al (2010) Pluripotency and cellular reprogramming: facts, hypotheses, unresolved issues. Cell 143(4):508–525

Hochedlinger K, Plath K (2009) Epigenetic reprogramming and induced pluripotency. Development 136(4):509–523

Ieda M et al (2010) Direct reprogramming of fibroblasts into functional cardiomyocytes by defined factors. Cell 142(3):375–386

Kim K et al (2010) Epigenetic memory in induced pluripotent stem cells. Nature 467(7313):285–290

Kloc M, Zagrodzinska B (2001) Chromatin elimination–an oddity or a common mechanism in differentiation and development? Differentiation 68(2–3):84–91

Kocaefe C et al (2010) Reprogramming of human umbilical cord stromal mesenchymal stem cells for myogenic differentiation and muscle repair. Stem Cell Rev 6(4):512–522

Lee G et al (2009) Modelling pathogenesis and treatment of familial dysautonomia using patient-specific iPSCs. Nature 461(7262):402–406

Lin Q et al (1997) Control of mouse cardiac morphogenesis and myogenesis by transcription factor MEF2C. Science 276(5317):1404–1407

Masip M et al (2010) Reprogramming with defined factors: from induced pluripotency to induced transdifferentiation. Mol Hum Reprod 16(11):856–868

Mattis VB, Svendsen CN (2011) Induced pluripotent stem cells: a new revolution for clinical neurology? Lancet Neurol 10(4):383–394

Meissner A (2010) Epigenetic modifications in pluripotent and differentiated cells. Nat Biotechnol 28(10):1079–1088

Nakatsuji N (2010) Banking human pluripotent stem cell lines for clinical application? J Dent Res 89(8):757–758

Osafune K et al (2008) Marked differences in differentiation propensity among human embryonic stem cell lines. Nat Biotechnol 26(3):313–315

Raya A et al (2009) Disease-corrected haematopoietic progenitors from Fanconi anaemia induced pluripotent stem cells. Nature 460(7251):53–59

Saha K, Jaenisch R (2009) Technical challenges in using human induced pluripotent stem cells to model disease. Cell Stem Cell 5(6):584–595

Swijnenburg RJ et al (2005) Embryonic stem cell immunogenicity increases upon differentiation after transplantation into ischemic myocardium. Circulation 112(9 Suppl):I166–I172

Tamaoki N et al (2010) Dental pulp cells for induced pluripotent stem cell banking. J Dent Res 89(8):773–778

Taylor CJ et al (2011) Immunological considerations for embryonic and induced pluripotent stem cell banking. Philos Trans R Soc Lond B Biol Sci 366(1575):2312–2322

Vierbuchen T et al (2010) Direct conversion of fibroblasts to functional neurons by defined factors. Nature 463(7284):1035–1041
Xie H et al (2004) Stepwise reprogramming of B cells into macrophages. Cell 117(5):663–676
Ye Z et al (2009) Human-induced pluripotent stem cells from blood cells of healthy donors and patients with acquired blood disorders. Blood 114(27):5473–5480
Zhao T et al (2011) Immunogenicity of induced pluripotent stem cells. Nature 474(7350):212–215
Zhou Q et al (2008) In vivo reprogramming of adult pancreatic exocrine cells to beta-cells. Nature 455(7213):627–632
Zhu H et al (2011) Investigating monogenic and complex diseases with pluripotent stem cells. Nat Rev Genet 12(4):266–275

Chapter 7
New Approach to Understand the Biology of Stem Cells

Colebrook (2002) claims that we are in a post-linguistic era and need to develop theories and approaches that are not language dependent. On the other hand, we have been dedicated firmly to use only language to describe extremely complex and interactive events in biology. If we think that mathematics is the language of nature then we should be able to represent and understand everything around us through numbers. When we turn those numbers of any system to the graphs, patterns emerge. Hence, there are patterns everywhere in nature (Aronofsky 1998). The way complex systems and patterns arise out of a multiplicity of relatively simple interactions is called emergence. Therefore, emergence is central to the theories of integrative levels and of complex systems (Corning 2002).

However, in twentieth century of the science, I assume that we are like colorblind humans in biology reminding very didactic story of Leon Lederman with Dick Teresi (1993) in their book called *The God Particle*. In this story some intelligent beings, who cannot see objects with sharp black and white, from imaginary planet Twilo come to the planet on a goodwill mission. To give them a test of our culture, hosts take them to World Cup soccer match as the most popular cultural event on the planet. Since they cannot see black-and-white, they watch the game with polite but confused looks on their faces. Lederman continues:

> As far as the Twiloans are concerned, a bunch of short-pantsed people are running up and down the field kicking their legs pointlessly in the air, banging into each other, and falling down. At times an official blows a whistle, a player runs to the sideline, stands there, and extends both his arms over his head while the other players watch him. Once in a great while the goalie inexplicably falls to the ground, a great cheer goes up, and one point is awarded to the opposite team.
>
> The Twiloans spend about 15 min being totally mystified. Then, to pass the time, they attempt to understand the game. Some use classification techniques. They deduce, partially because of the clothing, that there are two teams in conflict with one another. They chart the movements of the various players, discovering that each player appears to remain more or less within a certain geographical territory on the field. They discover that different players display different physical motions. The Twiloans, as humans would do, clarify their search for meaning in World Cup soccer by giving names to the different positions played by each footballer. The positions are categorized, compared, and contrasted. The

S. Yildirim, *Induced Pluripotent Stem Cells*, SpringerBriefs in Stem Cells
DOI: 10.1007/978-1-4614-2206-8_7, © The Author(s) 2012

qualities and limitations of each position are listed on a giant chart. A major break comes when the Twiioans discover that symmetry is at work. For each position on Team A, there is a counterpart position on Team B.

With two minutes remaining in the game, the Twiloans have composed dozens of charts, hundreds of tables and formulas, and scores of complicated rules about soccer matches. And though the rules might all be, in a limited way, correct, none would really capture the essence of the game. Then one young pipsqueak of a Twiloan, silent until now, speaks his mind. "Let's postulate," he ventures nervously, "the existence of an invisible ball."

"Say what?" reply the elder Twiloans.

While his elders were monitoring what appeared to be the core of the game, the comings and goings of the various players and the demarcations of the field, the pipsqueak was keeping his eyes peeled for rare events. And he found one. Immediately before the referee announced a score, and a split second before the crowd cheered wildly, the young Twiloan noticed the momentary appearance of a bulge in the back of the goal net. Soccer is a low-scoring game, so there were few bulges to observe, and each was very short-lived. Even so, there were enough events for the pipsqueak to note that the shape of each bulge was hemispherical. Hence his wild conclusion that the game of soccer is dependent upon the existence of an invisible ball (invisible, at least, to the Twiloans).

The rest of the contingents from Twilo listen to this theory and, weak as the empirical evidence is, after much arguing, they conclude that the youngster has a point. An elder statesman in the group—a physicist, it turns out—observes that a few rare events are sometimes more illuminating than a thousand mundane events. But the real clincher is the simple fact that there must be a ball. Posit the existence of a ball, which for some reason the Twiloans cannot see, and suddenly everything works. The game makes sense. Not only that, but all the theories, charts, and diagrams compiled over the past afternoon remain valid. The ball simply gives meaning to the rules.

This is an extended metaphor for many puzzles in physics, and it is especially relevant to particle physics. We cannot understand the rules (the laws of nature) without knowing the objects (the ball) and, without a belief in a logical set of laws we would never deduce the existence of all the particles" (Lederman and Teresi 1993).

We may have such an "alien" identity in unfathomably complex system of cell biology. In order to gain a "native" way, we need to think differently and look differently to see all colors and shapes. There might be several ways to do this. However, none could be as colorful and meaningful as what Benoit Mendelbroth brought to our sight. Mendelbroth took endless repetition of self-similarity for a lot of biological forms to exist e.g. tree; each of pates of branching is very similar. His seminal book, *The Fractal Geometry of Nature* published in 1982. This maverick mathematician passed away in his 85 leaving behind fractal geometry that was applied to physics, biology, finance and many other fields. His own description about himself is summarizing what he did for science (Hoffman 2010):

> I decided to go into fields where mathematicians would never go because the problems were badly stated. I have played a strange role that none of my students dare to take.

Looking at the picture below in Fig. 7.1, although it is black and white, I cannot help myself to think how much mathematical sense and foreseeing in biology we need. Would it be possible to engineer such a cell layer by layer? Would it be possible to explain all the events of all levels of that construction by the help of fundamental mathematics and physics?

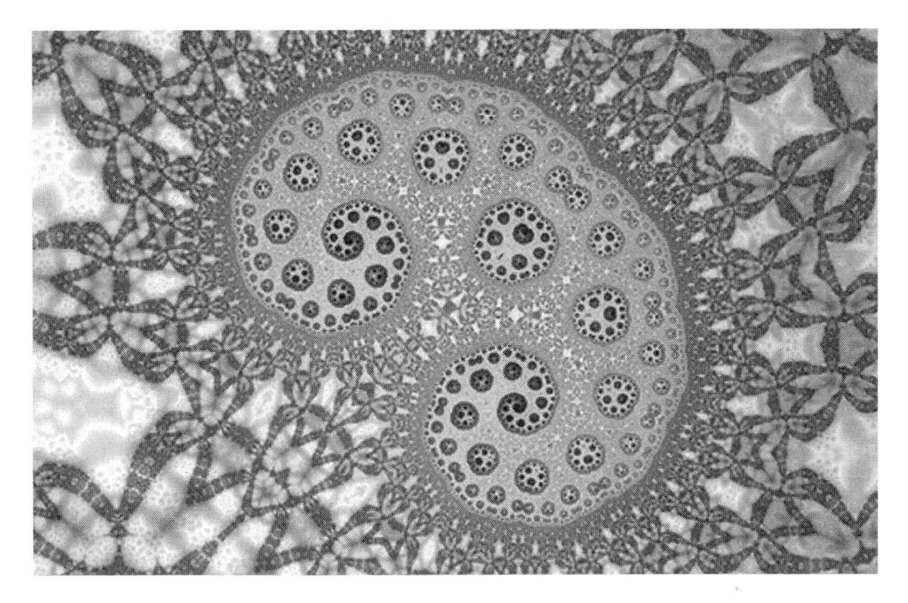

Fig. 7.1 The picture is a result of the fractal formula of $z = (z^*c + 1) + 1/(z^*c + 1)$ (Softologyblog 2011)

Goodwin (1996), in his book *How Leopard Changed Its Spots,* defines the unification of biology, physics and mathematics:

> The result is a unification of biology, physics, and mathematics that is accelerated by studies in the sciences of complexity and the realization that similar types of dynamic behavior arise from complex systems, irrespective of their material composition and dependent primarily on their relational order—the way the parts interact or are organized. Biology thus becomes more physical and mathematical, putting the insights of genetic, developmental, and evolutionary studies into more precise dynamical terms; at the same time, physics becomes more biological, more evolutionary, with descriptions of the emergence of the four fundamental forces during the earliest stages of the cosmic Big Bang, the growth sequences of stars, and the formation of the elements during stellar evolution. Instead of physics and biology remaining opposites, the former seen as the science of rational order deduced from fixed laws of nature and the latter described (since Darwin) as a historical science, physics is becoming more evolutionary and generative while biology is becoming more exact and rational (p. 171).

7.1 Health Versus Disease

Transplantation of corrected or differentiated iPSCs into diseased human tissue might be the most challenging issue for the prognosis of reprogramming technology. The ultimate and the most desired goal in regenerative medicine is to replace the dead or non-functioning cells that cause disease symptoms with healthy, laboratory-produced cells (Wilmut et al. 2011). We may obtain large quantities of cells and tissues from specific sets of patients; however, it is still unclear whether cells that came from reprogramming technology would reconstitute the tissues at the diseased sites is still unclear (Saha and Jaenisch 2009). Now we may think more deeply whether the human body is a machine where broken parts are replaceable with the spare parts.

The very contemporary insights are incorporated into evolutionary theory and complexity in Goodwin's (1996) mentioned book. He defines organisms as:

> Organisms are endowed with powerful particulars that give them the capacity to regenerate and reproduce their own natures under particular conditions, whereas inanimate systems cannot.
>
> This is an emergent property of life that is not explained by the properties of the molecules out of which organisms are made, for molecules do not have the capacity to make a whole from a part. DNA and RNA can make copies of themselves under particular conditions, but this is a self-copying process, not one in which a more complex whole is generated from a part. This is a principle reason organisms cannot be deduced to their genes or molecules. The particular type of organization that exists in the dynamic interplay of the molecular parts of an organism, which I have called a morphogenetic or a developmental field, is always engaged in making and remaking itself in life cycles and exploring its potential for generating new wholes.
>
> …
>
> We have now recovered organisms as the irreducible entities that are engaged in the process of generating forms and transforming them by means of their particular qualities of action and agency, or their causal powers. This includes hereditary particulars that give organisms a type of memory, and the intimate relations of dependence and influence between organisms and their environments. The life cycle includes genes, environmental influences, and the generative field in a single process that closes on itself and perpetuates its nature generation after generation. Species of organisms are therefore natural kinds, not the historical individuals of Darwinism. The members of a species express a particular nature (p 176–177).

Goodwin's statements parallel to German philosopher Immanuel Kant on the distinction between mechanisms and organisms are legendary:

> Kant described a mechanism as a functional unity in which the parts exist for one another in the performance of a particular function. … An organism, on the other hand, is a functional and a structural unity in which the parts exist for and by means of one another in the expression of a particular nature. This means that the parts of an organism—leaves, roots, flowers, limbs, eyes, heart, brain—are not made independently and then assembled, as in a machine, but arise as a result of interactions within the developing organism. … So organisms are not molecular machines. They are functional and structural unities resulting from a self-organizing, self-generating dynamic (p. 197).
>
> The emergent qualities that are expressed in biological form are directly linked to the nature of organisms as integrated wholes, which can be studied experimentally and stimulated by complex nonlinear models (p. 199).

Furthermore, Goodwin explains how our realization of a human being influence the way of treating the illness:

> If humans are to be understood essentially in terms of genes and their products, then illness is to be corrected by manipulating them. The result is drug-based medicine and genetic counseling or engineering. These can be extremely effective in certain circumstances, but medical care based on this approach focuses on illness rather than on health (p. 205).

Finally the writer cited Ingold's (1990) statement:

> Organisms and persons are not the effect of molecular and neuronal causes, of genes and traits, but instances of the unfolding of the total relational field. They are formed from relationships, which in their activities they create anew (p. 210).

Such novel insights need to exceed strong barriers in cell biologists who are very loyal to conventional theories of life. Fortunately, those ideas are coming with a simple mathematics and physics as much as possible.

7.2 Changed Paradigm: Reprogramming as Rare But Robust Process

During development, there is a gradual loss of differentiative potency proceeding from totipotency to pluripotency and multipotency in committed cell lineages toward terminal differentiation (Hemberger et al. 2009). The thought was broadly accepted and deeply rooted for many years in developmental biology that once a cell has terminally differentiated and become lineage committed, then it loses the potential of producing other cell types (Huang 2009).

In late 1950s Waddington introduced the term epigenetics to describe the unfolding of the genetic program for development. To Waddington, epigenetics was not very different from embryology, but it was a theory of development that proposed the early embryo was undifferentiated, and changed it to epigenetics (Waddington 1959). His epigenetic landscape is a metaphor to represent the way that developmental decisions are made. One common metaphor was a ball placed on a landscape, where the shape of the landscape "attracts" the ball more likely to follow certain channels and end up in certain places (Fig. 7.2). These lowest points represent the eventual cell fates, that is, tissue types. All the cells in the embryo would evolve according to the same laws, but because of the existence of inducing signals, cells in different regions would follow different pathways and end up at different attractors, which can be elegantly associated with different states of terminal differentiation. Once in its final valley, the ball cannot easily cross the mountain into neighboring valleys or return to the beginning (Waddington and Robertson 1966; Slack 2002).

Since the NT experiments have shown that, the nucleus of most, if not all, adult cells retains nuclear plasticity and can be rebooted to an embryonic state (Byrne et al. 2007), the recent groundbreaking reversion of this assumingly and potentially irreversible developmental process by the derivation of mouse iPSCs from adult dermal fibroblasts has surprised many cell biologists (Takahashi and Yamanaka 2006). While Waddington's epigenetic landscape metaphor is available to understand differentiation status of a cell, further explanations are needed to understand molecular nature and epigenetic barriers for reprogramming. It is obvious that the cell-fate determination and reprogramming are complex systems by emerging patterns from divergent genetic and epigenetic signals.

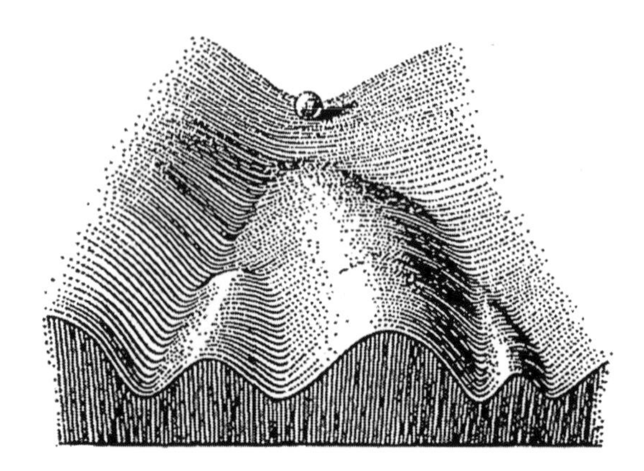

Fig. 7.2 In Waddington's landscape [from reference Waddington (1957)] ball represents a totipotent fertilized egg. It will differentiate into various lineages while the ball rolls down from the valley

7.3 From Reductionism to Wholeness

As Sui Huang (2011) stated clearly, molecular pathways, typically schematized in the form of an arrow–arrow diagram (A→B→C→etc.), represent biochemical cascades and are seen as the molecular embodiment of chains for many cell phenotypes. Hence, researchers preferred to use that arrow–arrow diagram in order to explain cell-fate decisions leading to describe many cell phenotypes. However, flood of genome-wide expression data along with many 'omics' data weakened this comfortable notion very much. This recent data set out the unfathomable complexity of molecular networks with their impenetrable density, therefore arrow–arrow schemes of reductive approaches of cell biology lost its clarity and simplicity (Huang and Wikswo 2006). Precluding the development of the concepts for understanding the collective parts is the biggest complication of the efforts to discover pluripotency and their molecular targets. The need for integrative approaches in the form of networks has started to be perceived among scientist in the middle of twentieth century (Kauffman 1969). Today *emergent* collective behaviors, e.g. pluripotency versus lineage commitment, can be explained by nested binary choices at the regulatory network-designed cell behavior level, which is far from simple linear causal explanation (Huang 2011).

Sui Huang, one of the pioneers of system biology is offering simple, generalizable patterns or principles that can be grasped by the human mind not withstanding the complexity of molecular interactions. This theoretical foundation is available to satisfy our intuitive comprehension by consistency with physical and mathematical principles (Huang 2011).

Huang (2004) tried to set up a pedagogical framework to describe sources of cellular states by the help of a complex high-dimensional integrated dynamical system. Huang, used natural and expected "the ground state" character of pluripotency to explain rarity and robustness of reprogramming events (Figs. 7.2, 7.3).

In order to understand the basics of dynamical system theory we need to get familiar with state spaces, attractors and regulatory networks. Therefore we may take the ball from Waddington's landscape and throw it into a bowl. Ball will

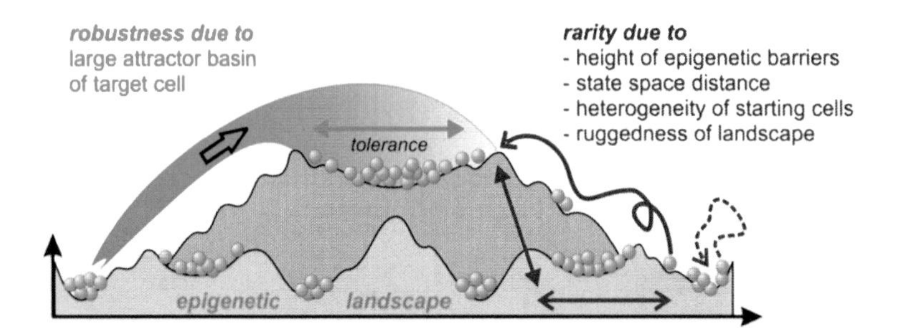

Fig. 7.3 Illustration of the co-existence of rarity and robustness in reprogramming the pluripotent state ("jumping back"). Note the "subattractors" ("wash-board potentials") as manifestation of the ruggedness of the epigenetic landscape, which imposes intermediate states that slow down the reprogramming events (with permission, from Huang 2009)

move around the bowl until it eventually comes to rest at the lowest point in the bowl. In dynamic systems that lowest point is called attractor (because ball was 'attracted' to that point). Without a very large perturbation the system becomes locked into a particular attractor. Now the whole ball is what dynamic system calls the basin of attraction of that system (Kauffman 1993). In cell biology, the basins are separated by some unstable states, which constitute the epigenetic barriers. Once an attractor is reached, the associated expression pattern is maintained (Huang 2009).

State space is simply an imaginary map of all the possibilities open to the system, e.g. for a coin toss it is just two points, heads or tails. In cell biology it is gene regulatory circuits. For a bistable gene regulatory circuit if, for instance, gene 1 (unconditionally) inhibits gene 2 or vice versa, there are only 2 possibilities: $X_1 \geq X_2$ or $X_1 \leq X_2$ (Fig. 7.4a). This particular 2-gene then generates two distinct attractor states: S_A has the expression pattern, $X_1 \geq X_2$, and S_B has the complementary pattern, $X_1 \leq X_2$. Since the two attractors can coexist within the same environmental conditions the circuit is said to be bistable. Attractor states are robust, "self-stabilizing", distinct states (Fig. 7.4b) (Huang 2009, 2011).

Huang (2011) summarizes the network dynamics as follows:

> An abstract network state $S(t) = (x_1, x_2, \ldots, x_n)$ at time t, which also represents a particular gene-expression profile, hence the state of a cell, is mapped into a point object characterized by its position.
>
> ...
>
> An entire network and its state S at a time t maps into one point in state space. The trajectory in state space captures the coordinated change in gene expression as dictated by the gene-regulatory network. Since the network state S also represents a gene-expression profile, which in turn determines the cell phenotype, a trajectory tracks the cell's phenotype change. The state space trajectory is thus a directed curve that truly represents a developmental process, such as differentiation. Unlike an arrow in a network diagram or a 'pathway' which is merely a shorthand symbol that has been over-interpreted as a causal explanation in biology, the arrow in state space or trajectory (Fig. 7.4c) is a formal physical entity and represents a biological process in its entireness; it is a true path.

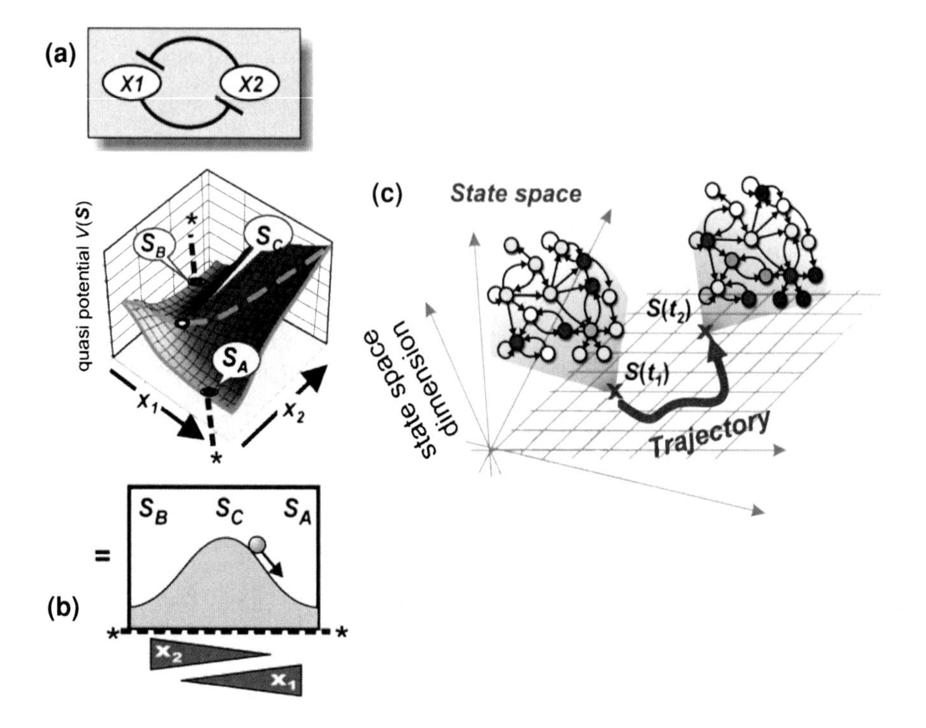

Fig. 7.4 The transformation from arrow–arrow approaches to system biology, which uses state spaces, trajectories and attractors. **a** Circuit architecture of two mutually inhibitory genes and examples of gene regulatory circuits using same binary decisions for cell differentiation in many multipotent cells. **b** *Dashed line* represents the separatrix, dividing the state space into the basins of attraction. *Bottom*: Simplified schematic representations, obtained from cross section along the *dashed line* *…*. **c** Any point in this space represents a (theoretical) network state S at time t, defined by the expression values x of the subnetwork's two genes, $S = (x_1, x_2)$ (gene-expression pattern) at time t. Since, as most states, they do not represent stable network states, they are driven by the network interactions to seek a stable state; hence they move in state space along trajectories (*solid line*) that lead to the stable attractor state. The trajectory represents the movement of the state that manifests the regulatory relationship 'X_1 inhibits X_2'—however, it is modulated by other inputs from the network. The states, S_1 *and* S_2, the perturbed trajectory all lie within the state-space region that 'drain' to the particular attractor S^*;. hence they all lie within its basin of attraction (with permission, from Huang 2009 and 2011)

Since all of the measurements about molecular networks are snapshots of a fixed moment, it has to ignore completely potentially functional variations. If we imagine a gene regulatory network as state of fluctuating gene expressions, even within homogenous cell populations, having a clear idea about the functional state of cells using mRNA expression values coming from broad range microarray analysis would not be ideal or realistic (Fig. 7.5).

If we go back to Huang's (2009) epigenetic landscape in Fig. 7.3, we are able to see that in the dynamic perspective, individual cells in a clonal population show fluctuations in their expression levels within the attractor basin and the cells that at

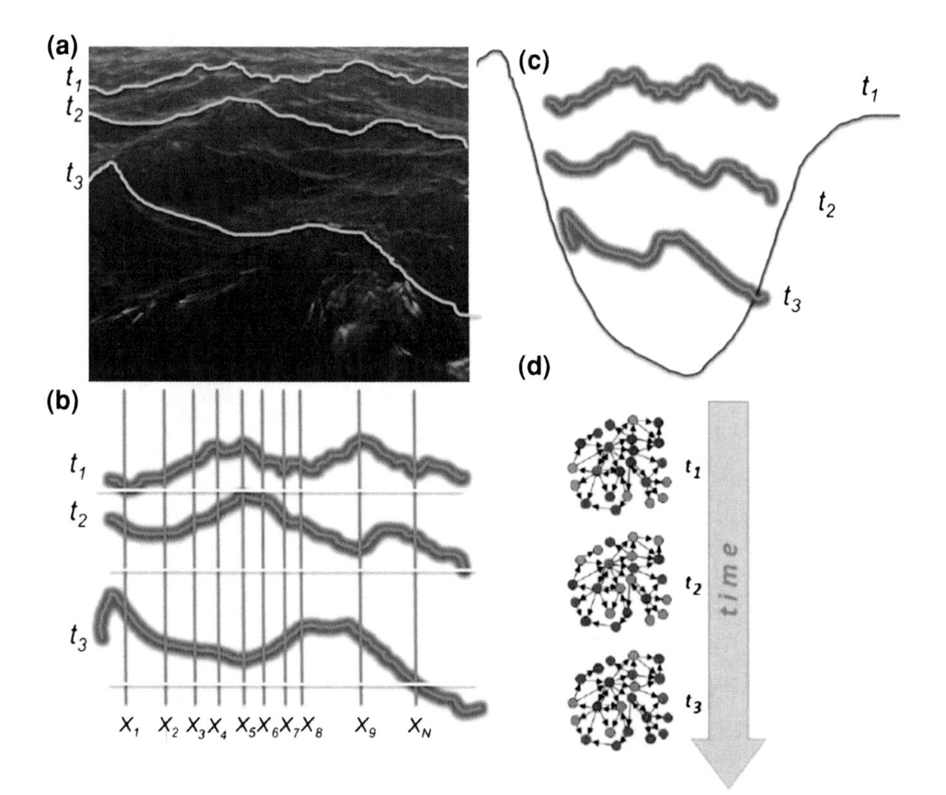

Fig. 7.5 Considering gene expressions as waves in a snapshot of a wavy sea (**a**), the expression of each individual gene X_1 to X_N would fluctuate in every consequent minute (**b**), although the phenotype of a cell is stable in a basin of allowed limits of fluctuations (**c**). Instead of describing interactions of genes in a arrow-to-arrow manner, a graphical representation of whole network would give a better idea through a dynamic system such as molecular network (**d**). (The graphical representation in d is displayed with the permission of Dr. Sui Huang)

the given time happen to be near the rim of the basin are most responsive to differentiating signals that kick them out of the stem cell attractor or destabilize the latter. Overall, the efforts for inducing differentiation pathways both in somatic or embryonic cells indicate that once the epigenetic barriers are exceeded by any reprogramming events, then the cell-fate changes (Hochedlinger and Plath 2009). If stem cells contain a heterogeneous mixture of microstates, each primed for a distinct fate, then transitioning into each other in a dynamic equilibrium (within the attractor basin) is highly possible when no fate-committing external cue is present (Huang 2009). Since Ying et al. (2008) presented ESCs have an innate program for self-replication that does not require extrinsic instruction; the ground state character of pluripotency can be accepted as natural default state. Then, with the help of Huang's (2009) complex high-dimensional dynamical system, we can understand why reprogramming pluripotency is robust yet rare:

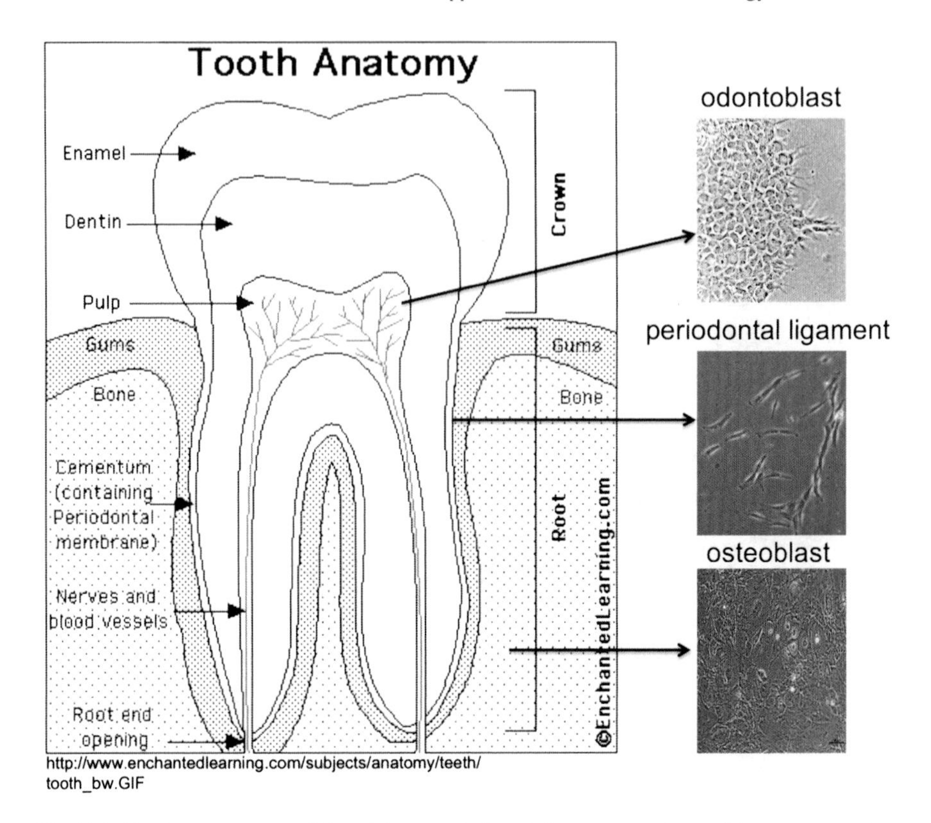

Fig. 7.6 Tooth components. Odontoblasts, periodontal ligaments and osteoblasts are secreting cells of three closely related tissues, dentin, periodontal membrane and alveolar bone

> Since the pluripotent state is an attractor state with a rather large basin of atraction, it is robust—a ground state.

And the rarity of reprogramming events can be explained because of the ruggedness of the attractor landscape:

> Only a small fraction of the cells in the population, namely those whose fluctuating microstate map into a gene expression pattern that fulfils some particular priming requirement, may actually be responsive to the nature of the reprogramming signals (Fig. 7.3) (Huang 2009).

7.4 More Future Considerations

It is obvious that differentiation and lineage commitment warrant a ground state model. If you could push ontogenically closed cells to a common place, mimicking differentiation pathways would follow the pathways through cellular specification

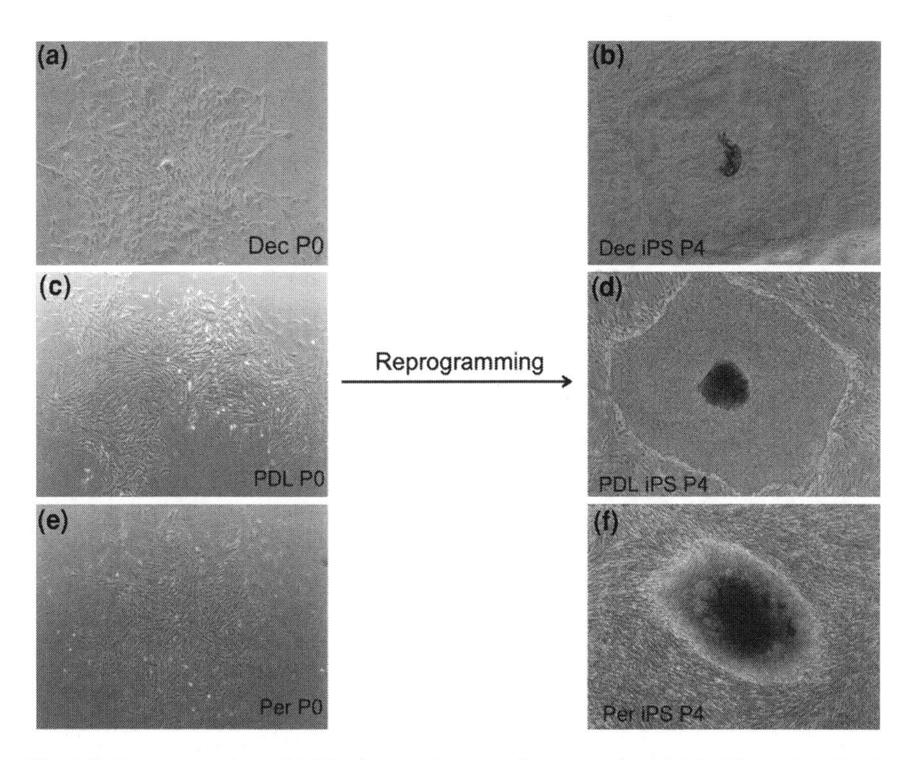

Fig. 7.7 Reprogramming of highly clonogenic stem cells derived from (**a**) deciduous dental pulp (Dec), (**c**) periodontal ligament (PDL) belong to a permanent tooth, and (**e**) dental pulp from same permanent tooth (Per) gave iPSC colonies, which are morphologically different (**b–f**)

aimed at the identification of the earliest lineage precursors. For example, tooth and alveolar bone duo may serve as a model to search such a close differentiation pathways, since there are very closely specialized tissues in an anatomical localization. Those tissues are alveolar bone of jaws, periodontal ligament, cementum and dentin of the tooth root, which are secreted by closely related cells: osteoblasts, periodontal ligament fibroblasts, cementoblasts and odontoblasts, respectively (Fig. 7.6).

They are all mesenchymal in origin except odontoblasts which have neuroectodermal origin (Koussoulakou et al. 2009). Although those tissues display many functional and physiological differences, there are no specific markers available to identify their specificity. On the other hand mesenchymal stem cell properties of periodontal ligament and dental pulp cells are almost identical (Huang et al. 2009). Dental pulp stem cells show clear difference than the other mesenchymal origined umbilical cord stromal stem cells (Oktar et al. 2011). However the attempt to reprogram them to iPS state showed varied phenotypes that can be moderated through passages (Fig. 7.7) (Yildirim, unpublished observation). It would be interesting to search if those morphological discrepancies reflect epigenetic status of different but closely related cellular origins, which arose.

References

Aronofsky D (1998) Pi. USA: 84 minutes.
Byrne JA et al (2007) Producing primate embryonic stem cells by somatic cell nuclear transfer. Nature 450(7169):497–502
Colebrook C (2002) Understanding Deleuze. Allen and Unwin, Crows Nest
Corning PA (2002) The re-emergence of "emergence": a venerable concept in search of a theory. Complexity 7(6):18–30
Goodwin B (1996) How the leopard changed its spots. Touchstone Book, New York
Hemberger M et al (2009) Epigenetic dynamics of stem cells and cell lineage commitment: digging Waddington's canal. Nat Rev Mol Cell Biol 10(8):526–537
Hochedlinger K, Plath K (2009) Epigenetic reprogramming and induced pluripotency. Development 136(4):509–523
Hoffman J (2010) Benoît Mandelbrot, Novel mathematician, Dies at 85. *New York Times* A28.
Huang S (2004) Back to the biology in systems biology: what can we learn from biomolecular networks? Brief Funct Genomic Proteomic 2(4):279–297
Huang S (2009) Reprogramming cell fates: reconciling rarity with robustness. Bioessays 31(5):546–560
Huang S (2011) Systems biology of stem cells: three useful perspectives to help overcome the paradigm of linear pathways. Philos Trans R Soc Lond B Biol Sci 366(1575):2247–2259
Huang S, Wikswo J (2006) Dimensions of systems biology. Rev Physiol Biochem Pharmacol 157:81–104
Huang GT et al (2009) Mesenchymal stem cells derived from dental tissues vs. those from other sources: their biology and role in regenerative medicine. J Dent Res 88(9):792–806
Ingold T (1990) An anthropologist looks at biology. Man (NS) 25:208–229
Kauffman S (1969) Homeostasis and differentiation in random genetic control networks. Nature 224(5215):177–178
Kauffman SA (1993) Self-organization and adaptation in complex system. Oxford University Press, New York
Koussoulakou DS et al (2009) A curriculum vitae of teeth: evolution, generation, regeneration. Int J Biol Sci 5(3):226–243
Lederman L, Teresi D (1993) The god particle if the universe Is the answer, what is the question?. Dell Publishing, New York
Oktar PA et al (2011) Continual expression throughout the cell cycle and downregulation upon adipogenic differentiation makes nucleostemin a vital human MSC proliferation marker. Stem Cell Rev 7(2):413–424
Saha K, Jaenisch R (2009) Technical challenges in using human induced pluripotent stem cells to model disease. Cell Stem Cell 5(6):584–595
Slack JM (2002) Conrad Hal Waddington: the last renaissance biologist? Nat Rev Genet 3(11):889–895
Softologyblog (2011) Archive for the 'Fractals' category. http://softologyblog.wordpress.com/category/fractals/
Takahashi K, Yamanaka S (2006) Induction of pluripotent stem cells from mouse embryonic and adult fibroblast cultures by defined factors. Cell 126(4):663–676
Waddington C (1957) The strategy of genes. George Allen & Unwin, London
Waddington CH (1959) Canalization of development and genetic assimilation of acquired characters. Nature 183(4676):1654–1655
Waddington CH, Robertson E (1966) Selection for developmental canalisation. Genet Res 7(3):303–312
Wilmut I et al (2011) The evolving biology of cell reprogramming. Philos Trans R Soc Lond B Biol Sci 366(1575):2183–2197
Ying QL et al (2008) The ground state of embryonic stem cell self-renewal. Nature 453(7194):519–523

Chapter 8
Conclusion

Reprogramming technology surprised many cell biologists at the beginning. IPSCs caused a real U turn in cell biology; cells can be reprogrammed into their embryonic states whenever you need. The techniques are evolved enormously fast and today there are tremendous investments to develop robust and sensitive ways to generate ideal pluripotent cells. However, for obtaining the pluripotent lines, their subsequent differentiation and characterization, validated techniques and reagents in order to achieve high quality and safe progenitor cells must be conducted under controlled conditions. While those processes are helping a lot to solve the biggest problem of medicine, they are also pushing minds to think more deeply about health vs. illness. Yet, iPSCs would have profound implications for both basic research and clinical therapeutics by providing a patient-specific model system to study the pathogenesis of disease and test the effectiveness of pharmacological agents, as well as by providing ample source of autologous cells that could be used for transplantation.

S. Yildirim, *Induced Pluripotent Stem Cells*, SpringerBriefs in Stem Cells

DOI: 10.1007/978-1-4614-2206-8_8,

About the Author

Sibel Yildirim is an associate professor at the Faculty of Dentistry, Department of Pediatric Dentistry, University of Selcuk, Konya, Turkey. She has been lecturer and practising dentistry at the university for years. She and her colleagues have established the first multi-disciplinary, dental/oral research centre in Turkey. She has authored many papers in international journals and has completed several research projects. Along with her pediatric dentistry PhD, she has also her second PhD on histology and embryology. Her research interests include the role of cytokines on deciduous tooth resorption, the viral ethiopathogenesis of pulpal/periapical diseases of deciduous teeth and vital pulpal therapies using recombinant human proteins or gene therapy. Accordingly, she has led some experiments on deciduous tooth pulp tissue and her team obtained important results showing high regulatory capacity of dental pulp cells on the events of deciduous tooth resorption or retention.

She had a chance to study with the indisputable leaders in the dental engineering field in Japan, Switzerland and the United States. She has become an expert in stem cells from the pulp tissue of deciduous and permanent teeth. She and her colleagues showed that dental pulp tissue stem cells have some clear differences than mesenchymal stem cells obtained from umbilical cord stroma. In addition to her dental tissue engineering studies she has pursued re-programming experiments in Geneva. In this part of her studies, she tried to generate the induced pluripotent stem cells (iPSCs) from deciduous vs. permanent dental pulp stem cells, as well as periodontal ligament cells and oral keratinocytes. She could observe that stem cells that have dental origin displayed strong potential for reprogramming. She believes comprehensively that isolating epithelial and mesenchymal stem cells from deciduous teeth and turning them to iPSCs will point toward novel venues for in situ restoration of dental tissue repair.

As a pediatric dentist, histologist and embryologist, she does believe that she has really powerful tool for understanding cell biology. After long years of experience in academic fields in three different continents she found herself that she became a student again who is keen to reach beyond the traditional boundaries

S. Yildirim, *Induced Pluripotent Stem Cells*, SpringerBriefs in Stem Cells
DOI: 10.1007/978-1-4614-2206-8, © The Author(s) 2012

of biology. After she read Sui Huang's and Stuart Kaufman's marvelous manuscripts, she convinced totally that biocomplexity is the only field that helps to achieve her goals; moving one step closer to understanding life.

Index

A
Attractor, 61–63

C
Complexity, 59, 60, 62, 68

D
Different pluripotent cells, 5, 6
Differentiation, 1, 2, 7, 24, 42
Disease in a dish, 33
Disease modeling, 42
Disease-specific iPS cells, 34, 40

E
Embryonic stem cells, 1, 5
ESC, 2, 8, 9, 11, 23, 27, 28, 33, 40, 43, 51–53
Epigenetic modifications, 26, 27, 54

F
Factor induced reprogramming, 21

G
Gene regulatory network, 43, 63, 64

I
Induced pluripotent stem cells, 1, 11, 15
iPSC, 51–53, 67
Integration free reprogramming, 14, 18, 30, 40, 41, 51, 53, 62, 65

P
Pluripotency, 2, 5–9, 11, 12, 21–24
Pluripotent cells, 2, 5–7, 9, 15, 25, 27, 28, 33, 53, 68, 71

R
Reprogamming, 23–28, 39, 40, 42, 51–53, 59–62, 65, 66, 68, 71

S
Similarities and differences between iPSCs and ESCs, 27
State space, 63, 64
Stem cells, 33, 39, 43, 53, 54, 57, 65, 67
Steps in reprogramming, 21
System biology, 62, 64

T
Transdifferentiation, 16, 53, 54

S. Yildirim, *Induced Pluripotent Stem Cells*, SpringerBriefs in Stem Cells
DOI: 10.1007/978-1-4614-2206-8,